ECOLOGICAL CLASSIFICATION OF SASKATCHEWAN'S MID-BOREAL ECOREGIONS USING RESOURCE MAPS AND AERIAL PHOTOGRAPHS

J.D. Beckingham,[1] V.A. Futoransky,[2] and I.G.W. Corns[3]

SPECIAL REPORT 14

Canadian Forest Service
Northern Forestry Centre
1999

[1, 2] Geographic Dynamics Corp., 10368B - 60 Avenue, Edmonton, AB T6H 1G9
[3] Canadian Forest Service, Northern Forestry Centre, 5320 - 122 Street, Edmonton, AB T6H 3S5

Catalogue no. Fo29-34/14-1999E
ISBN 0-660-17864-8
ISSN 1188-7419

This publication may be purchased from:

UBC Press
c/o Raincoast Fulfillment Services
8680 Cambie Street
Vancouver, British Columbia V6P 6M9
Phone: 1-800-561-8583, 604-323-7100
Fax: 1-800-565-3770
E-mail: custserv@raincoast.com

Printed in Canada

Canadian Cataloguing in Publication Data

Beckingham, John D. (John David), 1961-

Ecological classification of Saskatchewan's mid-boreal ecoregions using resource maps and aerial photographs

(Special report, ISSN 1188-7419; no 14)
Co-published by University of British Columbia Press.
Includes an abstract in French.
Includes bibliographical references.
ISBN 0-660-17864-8
Cat. no. Fo29-34/14-1999E

1. Forest ecology — Saskatchewan — Classification.
2. Forest site quality — Saskatchewan.
3. Aerial photography in forestry — Saskatchewan.
4. Ecological mapping — Saskatchewan.
5. Forest surveys — Saskatchewan.
I. University of British Columbia.
II. Futoransky, V.A.
III. Corns, I.G.W. (Ian George William)
IV. Northern Forestry Centre (Canada)
V. Series: Special report (Northern Forestry Centre (Canada)); no. 14.
VI. Title.

Beckingham, J.D.; Futoransky, V.A.; Corns, I.G.W. 1999. ***Ecological classification of Saskatchewan's mid-boreal ecoregions using resource maps and aerial photographs.*** **Nat. Resour. Can., Can. For. Serv., North. For. Cent., Edmonton, Alberta. Spec. Rep. 14**

ABSTRACT

Five air photo stereograms representing landscape profiles of a wide range of ecological site conditions (ecosites) in the boreal mixedwood ecoregions of Saskatchewan are presented. Procedures are described to identify and map mixedwood ecosites using five sources of existing information including: ecoregion maps, an ecological classification field guide, forest inventory maps, soil survey maps and reports, and aerial photographs, with dichotomous keys. Ecosites described in detail in the companion Saskatchewan ecosite classification field guide are characterized in this book by attributes described in these five resource information sources. Also included are a glossary of technical terms, and appendixes listing ecosites, ecosite phases, plant community types, and soil types of the Saskatchewan mixed-wood ecoregions, information on boreal tree silvics, soil mapping associations, and the correspondence of ecosites and forest inventory map units.

RÉSUMÉ

On présente cinq stéréogrammes aériens représentant les profils de paysage d'une large gamme d'états de sites écologiques (écosites) des écorégions boréales mixtes de la Saskatchewan. On y décrit les procédures d'identification et de cartographie des écosites mixtes à l'aide de cinq sources d'informations, incluant : des cartes des écorégions, un guide de terrain de classement écologique, des cartes d'inventaire forestier, des cartes et des rapports pédologiques et des photographies aériennes, avec des clefs dichotomiques. Les écosites décrits en détail dans le guide d'accompagnement de classification des écosites de la Saskatchewan y sont caractérisés par les attributs expliqués dans les cinq sources d'informations utilisées. On y trouve également un glossaire de termes techniques ainsi qu'une annexe donnant la liste des écosites, de leurs phases, des types de communautés végétales et des types de sols des écorégions mixtes de la Saskatchewan, des informations sur l'écologie forestière boréale, des associations pédologiques et la correspondance entre les écosites et les unités cartographiques d'inventaire forestier.

ACKNOWLEDGMENTS

This ecosite classification manual was developed with funding from the Canada–Saskatchewan Partnership Agreement in Forestry as a companion document to the *Field guide to ecosites of the mid-boreal ecoregions of Saskatchewan*, Special Report 6. The original direction of Philip Loseth, Project Facilitator, formerly of the Canadian Forest Service in Prince Albert, Saskatchewan is acknowledged.

We acknowledge Jamie Benson, Bob Reed, and Michael McLaughlan, R.P.F., from Saskatchewan Department of Environment and Resource Management who participated in the review process. Their valuable input is appreciated. Special thanks to David Lindenas and Don Ens, also from Saskatchewan Department of Environment and Resource Management, for providing forest inventory resources (maps and aerial photographs) used to develop this manual.

CONTENTS

APPENDIXES

FIGURES

TABLES

1.0 INTRODUCTION

Ecological classification has become an important component of current forest management practices. It provides an ecological basis for integrated resource management. The mapping of ecological units can be time-consuming and expensive; therefore, any information or tools that provide assistance in the classification and mapping of ecological units should be utilized to save time, reduce expenses, and increase accuracy. Aerial photograph interpretation, forest cover maps, and soil survey maps and reports provide valuable information about the distribution of ecological units across the landscape.

The *Field guide to ecosites of the mid-boreal ecoregions of Saskatchewan* (Beckingham et al. 1996) presents an ecological classification system that has been designed to organize ecological information and provide a structure for ecologically based forest management. This classification system is site-based and takes into consideration site, soil, and vegetation characteristics. In order to apply this type of classification system to an area and integrate the information with resource management systems, it is necessary to collect some basic ecological information. This can be accomplished by conducting field surveys; however, the field collection of ecological data can be very time-consuming and costly. It is helpful, therefore, to apply other sources of information when conducting ecological site classification.

Aerial photograph interpretation, in conjunction with the utilization of forest inventory and soil survey information, is one approach to implementing a general level of ecological classification. It should be used to complement field level investigations, not replace them. The results of this type of classification can be applied to many forest management practices, especially those requiring general or regional planning. This approach to ecological classification can also be useful for identifying areas that may require more detailed field surveying, thus balancing the required resolution of ecological classification with time/cost effectiveness.

This aerial photograph interpretation guide has been developed to provide insight in applying ecological information to classify and map forest land. The ecological information has been interpreted from aerial photographs, forest inventory maintenance maps, and soil survey maps. It is designed to be used in conjunction with Beckingham et al. (1996) and focuses only on those ecological units that are found within the classification system. This aerial photograph interpretation guide has been developed with the assumption that aerial photograph interpreters using the guide are knowledgeable and experienced in aerial photograph interpretation. Its primary purpose is to provide users with a tool for ecosite description and to illustrate how existing resource information can be used to understand the distribution of ecosites across the landscape. This aerial photograph interpretation guide is designed to promote thought and innovation during landscape interpretation. This guide includes:

- an overview of the classification system that is presented in Beckingham et al. (1996);
- a review of the types and sources of ecological information that can be applied when classifying an area using aerial photograph interpretation methods;
- a step-by-step application of the system to successfully utilize the above-mentioned sources with aerial photograph interpretation;
- keys and tables to help identify site types using aerial photograph interpretation methods;
- examples of typical toposequences of the mid-boreal ecoregions of Saskatchewan and how associated sites are classified using aerial photograph interpretation methods; and
- general descriptions of the ecology and identification of major tree species.

2.0 ECOLOGICAL CLASSIFICATION SYSTEM

Saskatchewan's ecosystem classification system has been developed as a means of organizing ecological information into a structured framework of ecological units that share similar environmental characteristics and respond to disturbance in a similar manner (Beckingham et al. 1996). As such, this system provides a basis for resource management and planning. Saskatchewan's ecosystem classification system consists of an integrated hierarchical ecological classification with three levels (ecosite, ecosite phase, plant community type) and a separate soil classification (Figure 1). The integrated hierarchical ecological classification and soil classification systems are nested within Saskatchewan's geographically based ecoregion classification system (Padbury and Acton 1994). A description of each level in the classification is provided below and the code and name of each ecological unit in the classification system is provided in Appendix 1.

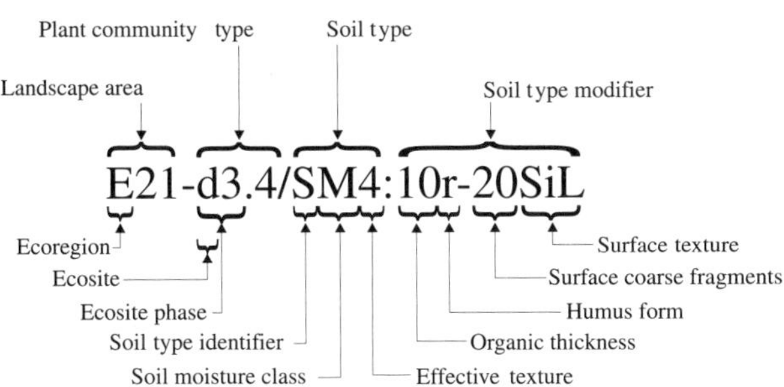

Figure 1. Mapping codes for the hierarchical ecological classification system.

Ecoregion (e.g., E Mid-Boreal Upland)

In Saskatchewan, an ecoregion can be thought of as "an area characterized by a distinctive regional climate as expressed by vegetation" (Lacate 1969). Reference ecosites are sites where the ecosystem is more strongly influenced by the regional climate than by edaphic (soil) or landscape factors. Reference sites are usually defined as having deep, well to moderately well drained soils of medium texture that neither lack nor have an abundance of soil nutrients or moisture, and are neither exposed to nor protected from climatic extremes (Strong and Leggat 1991). The term modal (Strong and Leggat 1991) and zonal (Pojar et al. 1991) have been used as alternative terms to describe reference sites. The current field guide was designed for use in the Mid-Boreal Upland and Mid-Boreal Lowland ecoregions of Saskatchewan (Beckingham et al. 1996). A capital letter is used to denote the ecoregion.

Landscape Area (e.g., E21 Montreal Lake Plain)

A landscape area is a subdivision of the ecoregion based on differences in physiography, surface expression, and the proportion and distribution of soils and plant community types within an area. A landscape area has a recurring pattern of slope, landform, soils, vegetation, and climate. A capital letter and a number are used to denote each landscape area (Padbury and Acton 1994).

Ecosite (e.g., d low-bush cranberry)

Ecosites are ecological units that develop under similar environmental influences (climate, moisture regime, and nutrient regime). They are groups of one or more ecosite phases that occur within the same portion of the edatope (moisture/nutrient grid) (Appendix 2). Each of the 13 ecosites are designated with a small letter, with "a" representing the driest ecosite and "m" the wettest. Each ecosite has been given a name that attempts to convey some information about the ecology of the unit. Ecosite, in this classification system, is a functional unit defined by moisture and nutrient regime. It is not tied to specific landforms or plant communities as in other systems (Lacate 1969), but is based on the combined interaction of biophysical factors, which together dictate the availability of moisture and nutrients for plant growth. Thus, ecosites differ in their moisture regime and/or nutrient regime.

Ecosite Phase (e.g., d3 low-bush cranberry tA-wS)

An ecosite phase is a subdivision of the ecosite based on the dominant species in the canopy. On lowland sites where a tree canopy may or may not be present, the tallest structural vegetation layer with a percent cover greater than 5 determines the ecosite phase. For example, the bog (j) ecosite has a treed and a shrubby ecosite phase. Generally, ecosite phases are believed to be mappable units and are the focus of this guide. They are identified by the ecosite letter (e.g., d) and name (e.g., low-bush cranberry) along with a number (e.g., 3) representing the phase within the ecosite (Figure 1).

Differences in ecosite phases of the same ecosite, although determined largely by the dominant species in the canopy, may be expressed as differences in lower strata plant species abundance and pedogenic processes. Ecosite phases, however, have a distinct range in canopy composition and lower strata floristics. The composition of the tree canopy as well as indicating environmental conditions (Lee 1924; Rowe 1956; Martin 1959; Mueller-Dombois 1964; Dix and Swan 1971; Carleton and Maycock 1981), influences structure, diversity, composition, and abundance of understory vegetation (Moss 1953; Rowe 1956; Dix and Swan 1971; Ellis 1986). The tree canopy and canopy-dependent factors such as degree of shading and understory species composition interact to dictate the type and quantity of organic matter, its rate of decomposition, and a site's nutrient availability. The ecosite phase level of the classification system, while being defined by canopy composition or structure, has a strong ecological basis. For forested sites the ecosite phase correlates well with forest cover types on provincial forest inventory maintenance maps.

Plant Community Type

(e.g., d3.4 tA-wS/low-bush cranberry–prickly rose)

Ecosite phases may be subdivided into plant community types, which are the lowest taxonomic unit in the classification system. The environmental characteristics of the community are defined at the ecosite level and to a lesser degree at the ecosite phase level. While plant community types of the same ecosite phase share vegetational similarities, they differ in their understory species composition and abundance. As one proceeds down the taxonomic levels of the hierarchy, from ecosite to ecosite phase to plant community type, the ecological variability within a unit is expected to decrease. Plant community types are considered difficult to map from aerial photographs; consequently, they will not be described further in this document.

Soil Type and Soil Type Modifier

Soil types are taxonomic units used to classify soils based on soil moisture regime, effective soil texture, organic matter thickness, and solum depth. Soil types can be used independently, in association with the hierarchical classification system (ecosite, ecosite phase, and plant community type), and to classify disturbed sites. The soil type is represented by a two- or three-character code (Appendix 3). When used with the hierarchical system, the soil type is separated from it with a slash (/) (Figure 1). The first letter in the code is an S (soil type identifier) followed by a capital V, D, M, W, R, or S, which represent very dry, dry, moist, wet, organic, and shallow soils, respectively. The third character of the soil type code, when required, is a number that represents the effective texture class in the very dry, dry, and moist soil types, or a letter that represents peaty (p) with the moist and wet soil types or mineral (m) with the wet soil types.

Soil type modifiers are used in association with the soil type as "open legend" modifiers to provide more resolution. Soil type modifiers include organic matter thickness, humus form, surface coarse fragment content, and surface texture.

An understanding of these concepts and a detailed description of individual ecological units can be obtained from Beckingham et al. (1996).

The resolution of the data that is available for classifying sites using aerial photograph interpretation methods allows for the classification system to be applied at the ecosite phase level only.

3.0 ECOLOGICAL INFORMATION: SOURCES

In order to apply the ecological classification system in a specific area, it is necessary to obtain ecological information about the area. A broad level of ecological unit mapping can be achieved through aerial photograph interpretation in conjunction with an evaluation and synthesis of existing published information. This, however, is not meant to replace but complement field evaluations. The main sources of information include aerial photographs, the *Field guide to ecosites of the mid-boreal ecoregions of Saskatchewan* (Beckingham et al. 1996), provincial forest inventory maintenance maps, and soil survey reports and maps. Table 1 outlines these sources of information, the agencies involved in their development, and a brief description of the information provided by each.

The user should not have to start from scratch when it comes to ecosystem mapping. A variety of information sources are available and can be easily used to gain an understanding of the ecology of the site to be described and classified.

Forest Inventory Maintenance Maps

Forest inventory maintenance maps are one of the tools that can be consulted to assist in classifying sites to the ecosite phase level. These maps display species composition, height, age, and density of a particular stand. In addition, estimates of soil drainage and texture are provided to further distinguish the site. All of these variables together provide a certain level of resolution to help classify a site. However, the forest inventory maintenance map forest cover types do not correlate directly with the ecosite phase forest cover types found in the field guide (Beckingham et al. 1996). Thus, this publication or the field guide should be consulted to ensure a proper match is obtained between forest cover types. Figure 2 outlines the sequence of map symbols used on the forest inventory maintenance maps and provides an example of a typical map symbol.

Table 1. Sources of land resource information for the mid-boreal ecoregions of Saskatchewan

Information type	Agency	Description
Ecoregion description		
Padbury and Acton (1994)	Ministry of Supply and Services Canada and Saskatchewan Property Management Corp.	1:2 000 000 scale map of the ecoregions of Saskatchewan.
Ecological classification		
Beckingham et al. (1996)	Canadian Forest Service	Ecological classification of the mid-boreal ecoregions of Saskatchewan. Includes vegetation, site, soil, and forest productivity information within 13 ecosites and 17 soil types.
Forestry		
Inventory maintenance maps	Saskatchewan Department of Environment and Resource Management, Forest Ecosystems Branch	1:12 500 scale UTM[a] series maps detailing stand data (species composition, height, age, density) and general site characteristics (soil drainage and soil texture class).
Soils		
Soil survey reports/maps	Saskatchewan Institute of Pedology, University of Saskatchewan	1:126 720 scale maps and associated reports including soil association, soil complex, and soil series.
Aerial photographs		
	Saskatchewan Department of Environment and Resource Management, Forest Ecosystems Branch	1:12 500 scale black and white aerial photographs.

[a]UTM = Universal Transverse Mercator.

General Designation: Height Class: Crown Closure Specific Designation	SH20D wS tA
Drainage Texture	WD-MWD MC-MF

Figure 2. Sequence of map symbols found on forest inventory maintenance maps (Lindenas 1985).

Map Unit/Unconformity (where present): Texture	Kk1/T: ls–sl
Landform: Slope Class: Stones (where present)	Fa3–4: St4–5

Figure 3. Sequence of map symbols found on soil survey maps (e.g., Ayres et al. 1978).

Soil Survey Reports/Maps

Soil survey reports provide regional soil information at a smaller scale of resolution (1:126 720) than the forest inventory. The soil survey maps delineate soil associations and soil complexes with descriptions of the common soil series within these soil map units. General information about the soil association parent materials, common soil development, soil texture, stone content, and slope classes are provided along with landform information. Figure 3 outlines the sequence of map symbols used on the soil survey maps and an example of a typical map symbol.

4.0 INTERPRETATION OF RESOURCE INFORMATION IN ECOLOGICAL CLASSIFICATION

A minimum amount of ecological data is required before an area of land can be classified to the ecosite phase level. The determination of edatope position (moisture regime and nutrient regime) and canopy composition are the minimum data required in order to carry out such a classification. However, the evaluation of other primary ecological variables is necessary in order to determine edatope position and canopy composition. Figure 4 shows how information from a variety of sources can be used to determine canopy composition, moisture regime, nutrient regime, and ultimately, ecosite phase. The following sections outline these data, their sources, and their application to ecological classification.

Tree species are the most conspicuous of all the vegetation layers. They provide managers with a basis for evaluating and classifying large land areas. In the boreal forest, tree species tend to occupy specific habitats, thus providing the interpreter with indications of site conditions.

This aerial photograph interpretation guide focuses on the ecosite phase level within the hierarchical ecological classification system. This is because the ecosite phase, as well as taking on the moisture and nutrient regime values of the ecosite to which it belongs, is defined by the species composition of the most conspicuous vegetation layer.

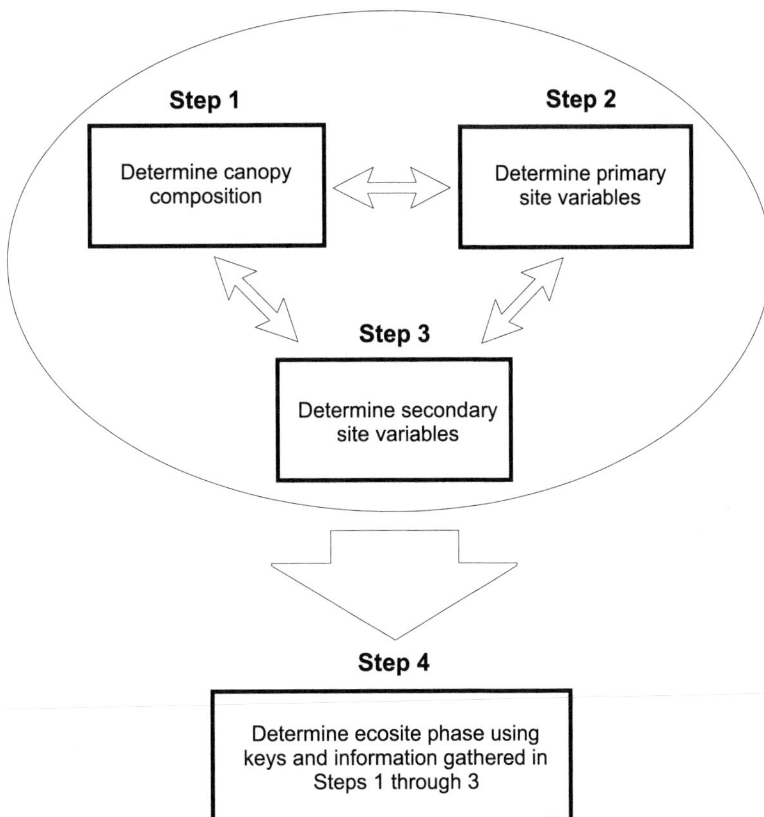

Figure 4. Steps involved to determine ecosite phase from aerial photographs and associated sources.

4.1 Ecological equivalence

The idea of ecological equivalence means that even though two sites are different in specific environmental factors, these factors act together so the sites are essentially equal in ecological function. Thus, the compensating factors together add up to the same ecologic sum. Differences in soil can add to or subtract from the apparent suitability or quality of any one category of topography percent slope, slope position, convexity or concavity, aspect, etc.) (Daubenmire 1976).

For example, an ecosystem may develop on a south facing slope that is similar to a stand that develops at lower altitude or latitude due to the compensating influence of topography on the local climate. As well, a similar ecosystem may develop along a slope in a seepage area, which could develop along a moist, nutrient-rich river floodplain due to the compensating influence of nutrient-rich discharge water.

Identical soil profiles may be associated with marked differences in productivity, whereas different sets of soil conditions can add up to the same ecologic sum (Daubenmire 1976). This problem stems from the lack of knowledge about the chemical and physical properties of the soil and their quantitative expression.

The spatial heterogeneity of soil properties in a complex landscape also add to the complexity of the assessment of productivity/soil relationships and its application to natural resource management. Analysis of soils has shown that important discontinuities in stable vegetation are often determined by environmental differences either too subtle or too ephemeral to evaluate with present pedologic technology (Viro 1961; Daubenmire 1973). On the other hand, stands that are closely similar in structure and floristic composition can occur on soils with very different profile characteristics (Daubenmire 1968). As plant succession advances, the degree of similarity of vegetation among ecologically equivalent sites increases and is at a maximum when competitive elimination has brought about a relatively stable condition (Daubenmire 1976).

4.2 Step 1 Determine canopy composition

Eleven tree species are common within the mid-boreal ecoregions of Saskatchewan. Eight occur in both the Mid-Boreal Upland (MBU) and Mid-Boreal Lowland (MBL) ecoregions: white spruce (*Picea glauca*), black spruce (*Picea mariana*), jack pine (*Pinus banksiana*), balsam fir (*Abies balsamea*), tamarack (*Larix laricina*), aspen (*Populus tremuloides*), balsam poplar (*Populus balsamifera*), and white birch (*Betula papyrifera*). White elm (*Ulmus americana*), green ash (*Fraxinus pennsylvanica*), and Manitoba maple (*Acer negundo*) are found only in the MBL ecoregion. The determination of canopy composition is critical to the successful classification of an area at the ecosite phase level. The tree species composing the canopy provide evidence as to site characteristics (moisture and nutrients) as well as being implicit in the classification of the ecosite phase. Specific guidelines for determining canopy composition in Saskatchewan are found in *Specifications for the interpretation and mapping of aerial photographs in the forest inventory section* (Lindenas 1985).

The following subsections provide information about the identification of tree species from aerial photographs along with information on how tree species can be used as indicators of physical site conditions. The identification of tree species and an understanding of the ecological significance associated with them is very important to the successful classification of land areas using aerial photograph interpretation methods.

4.2.1 Tree identification from aerial photographs

Tree species can be identified through aerial photograph interpretation by experienced interpreters or obtained directly from forest inventory maintenance maps. Forest inventory maintenance maps were developed through the interpretation of aerial photographs and provide information about tree species composition, height and crown closure. It is important to be able to recognize tree species from aerial photographs especially when forest inventory maintenance maps are not available for the area or when they require updating.

Recognition of tree species comes from an understanding of crown characteristics, texture, and photograph tone (Gimbarzevsky 1973). Crown characteristics that are important include tone, texture, shape, shadow, size, density, and distinctness. Individually, these characteristics may not lead to species recognition, but when considered in combination, they can provide strong evidence. Three general steps are commonly accepted for species identification (Sayn-Wittgenstein 1960; Howard 1970; Avery 1978):

- eliminate species that will not occur in the given area on the basis of environmental factors;
- apply knowledge of species associations to determine which species might be present; and
- examine crown characteristics.

Often it is difficult to distinguish between tree species on aerial photographs. Successful identification is dependent on the quality of the photographs, scale, season of photography, type of film, the interpreter's experience, and the interpreter's knowledge of the ecology of the species and the area. For example, black spruce and white spruce are difficult to separate on aerial photographs because their tone, texture, and shape often appear very similar. Areas where a high degree of uncertainty is associated with species identification should be considered for ground survey. Aerial photograph interpretation techniques supplemented by ground truthing ensures adequate levels of accuracy in site classification (van Kesteren 1992).

Silhouettes and descriptions of each tree species including ecological characteristics and a description about their typical appearances on aerial photographs is provided in Appendix 4. Tree species identification tips are also included.

4.2.2 Trees as indicators

In the mid-boreal ecoregions of Saskatchewan, the tree canopy is the most conspicuous of the vegetation layers. It is an important indicator of site conditions, influences understory species composition and abundance, and affects the availability and cycling of nutrients on a site. Dix and Swan (1971) and Carleton and Maycock (1981) both suggested that species occurrences, associations, and co-occurrences are the result of edaphic factors and substratum moisture regimes.

Jack pine stands are frequently associated with dry, sandy soils, aspen with loamy soils and intermediate moisture status, and black spruce with moist soils (Lee 1924; Rowe 1956; Martin 1959; Mueller-Dombois 1964; Dix and Swan 1971). Balsam poplar and black spruce are commonly located on poorly drained sites while aspen and white spruce favor intermediate moisture conditions (Dix and Swan 1971). Carleton and Maycock (1981) stated that certain canopy types such as tamarack and balsam poplar occur on sites with noticeable imports of nutrients in the form of lateral subsurface flow or seasonal flooding and sediment deposition. This is also true of the green ash, white elm, and Manitoba maple, which are commonly found on the floodplains of the MBL ecoregion. These sites support richer floras than jack pine and black spruce dominated stands, which are associated with tight ecosystem nutrient cycles (Carleton and Maycock 1981).

White birch and balsam poplar appear to be associated with a narrow range of pH (below 5.3 for white birch and above 7.2 for balsam poplar), and the other tree species tolerate a wide range of pH (Dix and Swan 1971). Dix and Swan also report that balsam poplar was the only species associated with clay textured soils. This is certainly not the case throughout Saskatchewan's boreal forest where a considerable proportion of the parent materials are fine textured and support a diversity of tree species. However, balsam poplar may have higher preference for clay soils than the other tree species of the boreal forest. In addition to indicating site conditions, canopy characteristics of trees have value in determining understory species composition and abundance.

Shading also has a dominant influence on the structure, diversity, and composition of understory vegetation. There is general consensus that as succession proceeds, canopy closure increases, which reduces light and increases shading in the understory (e.g., Rowe 1956; Ellis 1986). This occurs through increases in canopy cover of the existing dominant and codominant tree species or through shifts in canopy dominance. As overstory crown cover increases, there is an inverse trend in the understory with a decline in density, diversity, and production (e.g., Pase 1958; Hedrick 1975; Hall 1978; Uresk and Severson 1989). As well, structural changes favor shorter growth forms (Rowe 1956).

Tree canopy dominance can have a significant influence on the amount of shading in the understory. Ellis (1986) concluded that the single most important factor influencing succession in *Populus tremuloides–Picea glauca* mixedwood forest understories in northern Alberta is the shift in tree canopy dominance from aspen (*Populus tremuloides*) to white spruce (*Picea glauca*). He found that species richness and diversity were highest during early stages of succession, when shade-tolerant and -intolerant species were present. They gradually declined as shade-intolerant species were eliminated with increasing evergreen overstory cover. He indicated, however, that early development (first 25 years) was not studied.

Ellis (1986) also found that increasing evergreen canopy cover correlated significantly with increasing stand age, a decline in total understory plant cover, a decline in understory vascular species richness, a decrease in summer-green shrub and forb cover, and an increase in understory evergreen cover. One of the most prominent effects of the shift from summer-green to evergreen canopy is the dramatic rise in understory moss cover (Moss 1953; Rowe 1956; Dix and Swan 1971; Ellis 1986). A reduction and change in light quality that occurs with a shift from deciduous to coniferous crown cover dominance is associated with many understory vegetation changes.

In the boreal mixedwood of Saskatchewan, Dix and Swan (1971) found that once tree crowns attained a height of 25 ft. or more, they could be arranged in order of increasing shade tolerance as jack pine, aspen, balsam poplar, white birch, black spruce, and balsam fir, respectively. White spruce varies greatly in the shade it casts. This may in part be due to the variable age structure on many white spruce stands in the boreal forest.

While wetland sites may or may not have a tree canopy, the presence, absence, and composition of plant species on a site provide important information about environmental and ecological conditions.

4.3 Step 2 Determine primary site variables

Primary site variables are ecologically important variables that can be inferred directly or indirectly through the interpretation of aerial photographs or resource maps. Primary site variables that should be evaluated include topographic position, soil drainage, slope, aspect, landform/parent material, soil texture, and soil association. In combination, the primary site variable can be evaluated to predict moisture and nutrient regime (secondary variables).

4.3.1 Topographic position

The topographic position of a given site is defined by its location in the landscape. Typical topographic position classes include crest, upper slope, middle slope, lower slope, toe, depression, and level (Figure 5). Topographic position is one factor that must be considered in the determination of drainage, moisture regime, and nutrient regime.

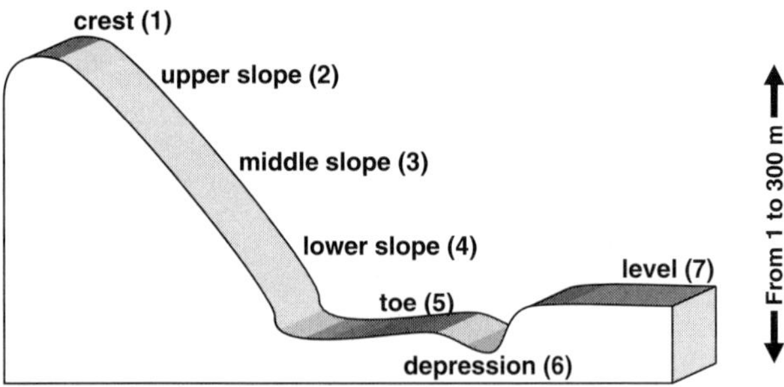

Figure 5. Schematic depiction of topographic position classes.

Topographic position is an important variable that can strongly influence soil moisture, soil nutrients and vegetation composition and abundance. Upper slopes receive moisture from precipitation only whereas lower slope positions receive both precipitation and ground water seepage. Seepage waters may be laden with nutrients that were acquired during percolation and subsurface flow. This increased moisture and nutrient availability results in very different species assemblages, including trees, because plants have very specific moisture and nutrient requirements. Jack pine, aspen, and white spruce are typically located in upper to midslope positions while balsam poplar, black spruce, and tamarack dominate lower slope and depressional sites. Stands on level topographic positions host a variety of ecosite types depending on the depth to the groundwater table and the texture of the parent material.

Topographic position can be readily determined from stereoscopic aerial photograph interpretation.

4.3.2 Soil drainage

Soil drainage could be thought of as a secondary site variable because primary site variables such as topographic position, slope, aspect, and texture are considered in its determination. Knowledge of topographic position and soil texture can be useful in determining soil drainage from aerial photographs (Table 2). Soil drainage, however, is treated in this guide as a primary site variable as it is used to understand moisture regime. Soil drainage is closely related to moisture regime and, therefore, is very important in its assessment. Soil drainage is the rate at which water is removed from the soil in relation to its supply. It is controlled by soil permeability, amount of water, and resistance of soil and parent material (Gimbarzevsky 1973). Drainage is classified into the following seven classes: very rapidly drained (VR), rapidly drained (R), well drained (W), moderately well drained (MW), imperfectly drained (I), poorly drained (P), and very poorly drained (VP).

Trees such as white spruce and aspen are relatively sensitive to poorly drained, cold, wet soils (Lutz and Caporaso 1959), thus the possibility of these species occurring on such soils is reduced. Soil drainage also helps us to understand the differences between very similar ecosites such as low-bush

Table 2. Prediction of soil drainage through topographic position and soil texture (*modified after* Lindenas 1985)

Soil texture	Topographic position					
	1	2	3	4, 5	6	7a	7b
Fine	W	W	W(MW)	MW–I	P–VP	W–MW	MW–P
Moderately fine	W	W	W(MW)	MW–I	P–VP	W	MW–P
Moderately coarse	W	W	W	MW(W)	P–VP	W	MW–P
Coarse	VR–R	VR–R	VR–R	R–MW	P–VP	VR–R	MW–P

VR very rapidly drained
W well drained
R rapidly drained
MW moderately well drained
I imperfectly drained
P poorly drained
VP very poorly drained

1. crest
2. upper slope
3. middle slope
4. lower slope
5. toe
6. depressional level
7a. level (upland)
7b. level (lowland)

cranberry (d) and dogwood (e) in terms of productivity and species composition. The dogwood (e) ecosite tends to have slightly poorer soil drainage and more seepage indicated by more robust trees, with a greater occurrence of balsam poplar and white birch.

Information about soil drainage is primarily given by the forest inventory maintenance maps. This information should be confirmed by an assessment of the interaction of topographic position interpreted from aerial photographs and soil texture from the soil survey maps. According to Gimbarzevsky (1973), climatic factors of the area must also be considered when applying soil drainage values in evaluation of land conditions.

4.3.3 Slope

The slope gradient of a site is the proportion of vertical rise relative to horizontal distance and is commonly expressed as a percentage. Zero degrees and zero percent slope describe a level site and 45° is equivalent to 100% slope. Slope is another primary site variable that should be considered in the determination of moisture and nutrient regimes. The orientation of the slope is the aspect (Section 4.3.4).

The greater the slope, the faster water will be removed from the site. Rapidly drained soils due to steep slopes are frequent in ridged and hummocky to steeply rolling terrain. Lichen (a) and blueberry (b) ecosites are common on rapidly drained sites. The steeper and longer the slope, the greater the hazard for soil erosion.

Slope gradient can be determined through aerial photograph interpretation by experienced interpreters. Soil survey maps have generalized slope gradients that are common for a delineated soil unit. The slope classes presented on the soil survey maps are as follows: 0–0.5%, 0.6–2%, 3–5%, 6–9%, 10–15%, and 16–30%.

4.3.4 Aspect

The aspect of a site is the orientation of the slope exposure indicated by a compass direction. Aspect provides an indication of the amount of solar radiation that is received by a site. When considered in combination with slope, topographic position and soil texture, it is valuable in determining moisture regime.

Steep, south-facing slopes have increased levels of insolation and tend to have less soil moisture available to plants. Jack pine and aspen are characteristic species found on these site types.

Aspect can readily be determined from aerial photographs. A level site has no aspect.

4.3.5 Parent material and soil texture

Parent materials are the deposits resulting from various landscape creation and destruction processes. The process of erosion and/or deposition is facilitated by water, wind, ice, gravity, or *in situ* degradation of the geological material.

Soil texture describes the proportions of sand, silt, and clay that make up the soil or surficial materials. Soil texture is essential in determining the moisture regime of a site (Beckingham et al. 1996, Section 5.7.2) and is generally linked closely with parent material and landform.

Some ecosites will only develop on very specific parent materials while others are less discriminating. For example, the Labrador tea–submesic (c) ecosite and the ostrich fern (f) ecosite tend to occur on morainal and fluvial deposits, respectively but can occur elsewhere if other conditions are suitable (see Section 4.1 on ecological equivalence). Soil types are derived from soil texture and moisture regime; therefore, soil types can be predicted if this information is known.

General soil texture information is given by soil survey maps and the forest inventory maintenance maps. In addition, parent material and landform information provided by the soil survey maps also provides information on soil texture characteristics, coarse fragment distribution, and other soil related attributes. For a list of parent materials referred to in this guide, see Table 3, and for texture classes, see Table 4 and Appendix 5.

4.3.6 Soil association

A soil association is a group of related soil series developed from similar parent material and occurring under essentially similar climatic conditions (Ayres et al. 1978). Soil associations are generally named after the location where they were first encountered.

Table 3. **List of parent materials and associated codes**

Code	Parent material
Mineral	
C	Colluvium
E	Eolian
F	Fluvial
GF	Glaciofluvial
GL	Glaciolacustrine
L	Lacustrine
M	Moraine/Till
Organic	
O	Organic (undifferentiated)
S	Swamp

Table 4. List of general texture classes

Texture class	Description
C	Coarse
MC	Moderately coarse
M	Medium
MF	Moderately fine
F	Fine

Most ecosite phases are found over a wide variety of soil associations and do not occur on one or the other exclusively. Trends, however, are evident with some ecosites. For example, the lichen (a) ecosite is almost exclusively found on the Pine Association (Pn). Some soil associations are limited to a small geographic area such as the Sipanok Association located in the MBL ecoregion where the ostrich fern mM-wE-bP-gA (f2) ecosite phase dominates.

Soil association information is obtained from soil survey reports and associated maps. See Appendixes 6 and 7 for soil associations of Saskatchewan's soil survey and for great groups of the Canadian System of Soil Classification.

4.4 Step 3 Determine secondary site variables

Secondary site variables are those ecologically important characteristics that must be interpreted from an evaluation of some or all of the primary site variables. The secondary site variables required to determine ecosite phase are moisture regime and nutrient regime.

4.4.1 Moisture regime

Moisture regime is a central concept of the ecological classification system and represents the moisture available for plant growth. Moisture regime is assessed through an integration of indicator plant species information and soil and site characteristics. Moisture regime is used to position the site on the vertical axis of the edatope (moisture/nutrient grid) and is essential in determining ecosite phase (Appendix 1). Beckingham et al. (1996, Section 5.7.2, Table 3) define moisture regime, moisture regime classes, and the characteristics of the moisture regime classes. Moisture regime has been divided into the following classes: very xeric (1), xeric (2), subxeric (3), submesic (4), mesic (5), subhygric (6), hygric (7), subhydric (8), and hydric (9).

Moisture regime can not be directly obtained from any of the information sources presented in this guide but must be inferred from available information in combination with ecological knowledge. Topographic position, soil drainage, slope, aspect, landform/parent material, and soil texture are all components of moisture regime (Figure 6).

4.4.2 Nutrient regime

Nutrient regime is an indication, on a relative scale, of the nutrient supply available for plant growth. Nutrient regime is expressed in classes: very poor (A), poor (B), medium (C), rich (D), and very rich (E), and its assessment is critical to ecological classification. The determination of nutrient regime involves inferences made on the basis of many information types, including several environmental and biotic components, and requires an understanding of ecology. Beckingham et al. (1996, Section 5.7.3) outline the factors that should be considered when assessing the nutrient regime of a site including: humus form, A horizon characteristics, soil texture, soil depth, coarse fragments, pH of parent material, seepage, and groundwater. Most of these variables can not be interpreted from aerial photography, but can help in developing an understanding of the specific nutrient status of a site.

Information about many of these categories can be extracted from the soil survey reports and maps, however, this information is relatively general and caution should be used when it is applied to a specific site. The interpreter, therefore, must apply knowledge of ecology, familiarity with the area, and draw from as many information sources as possible, including the ecological implications associated with the presence/absence of tree species.

4.5 Step 4 Determine ecosite phase

Several tools are provided in order to determine ecosite phase. Typically, once available information about a site is gathered a general picture begins to form. The information will help reduce the number of choices and eventually lead to the site's correct classification. Refer to the Ecoregions of Saskatchewan map (Padbury and Acton 1994) to determine the site location with respect to ecoregion boundaries. This may help the decision-making process by ruling out non-typical ecosite phases for that area.

Section 5.0 of this guide illustrates tools, in the form of keys and tables, to help determine ecosite phase. Remember, try to integrate information from all possible sources including aerial photographs, forest inventory maintenance maps, and soil survey reports and maps. Cross-reference your ecosite classification with the field guide by Beckingham et al. (1996) for further details about the site. If your ecosite phase choice does not match, then trace your steps and use the

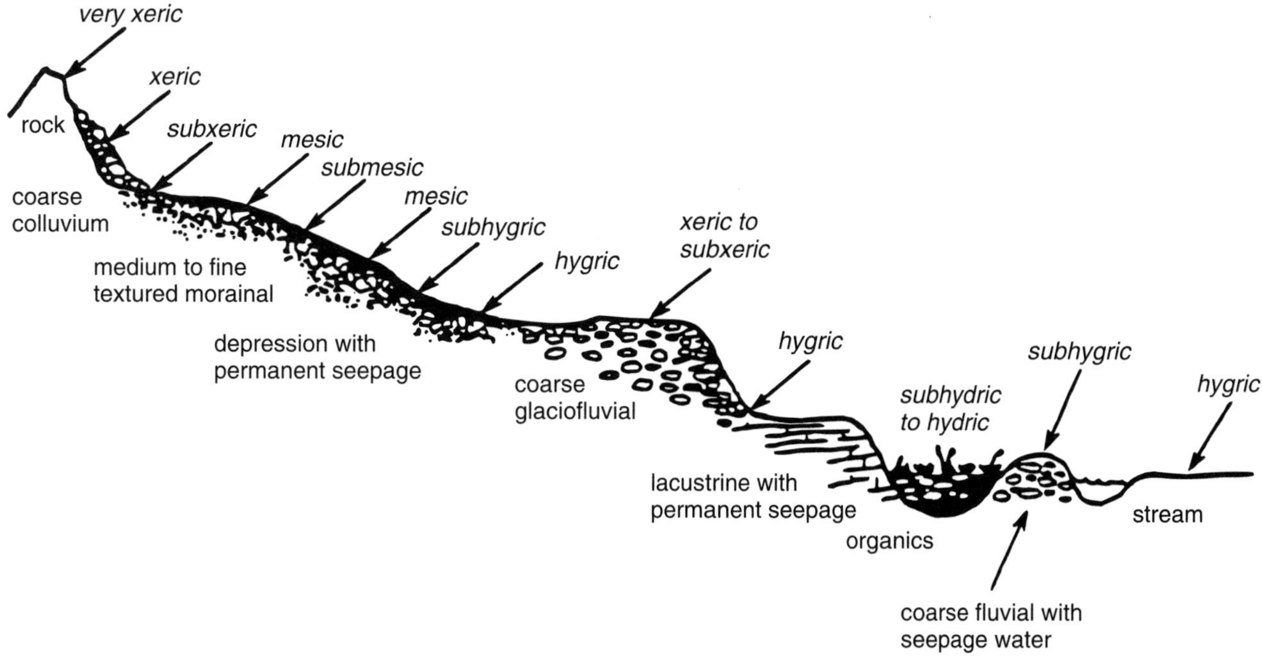

Figure 6. Moisture regime in relation to topographic position and parent material (*after* Alberta Forestry, Lands and Wildlife 1991).

keys and tables to determine a better ecosite phase classification. If you have what might be considered a transitional or ecotonal site then, based on the site conditions, choose the ecosite where the probability of improper management is minimal. The user must consult the management interpretation bars within the field guide (Beckingham et al. 1996) and determine how the site is affected by potential impacts. By comparing and selecting the ecosite phase with the highest potential for impact, the user is essentially managing that site for the worst-case scenerio, thus reducing the chance of negative impacts on the site.

If a site has been recently disturbed by fire, logging, or development you may not be able to classify it to the ecosite phase level. In these cases, when the canopy is missing or the vegetation is disturbed, the ecosite level becomes the appropriate level of classification.

5.0 KEYS TO IDENTIFYING ECOSITE PHASE THROUGH THE INTERPRETATION OF RESOURCE INFORMATION

This section provides information in the form of keys and tables that assist the user in the identification of ecosite phases from aerial photographs and existing resource information. The user must extract information from all sources, and in conjunction with any previous ecological knowledge and aerial photograph interpretation experience, use the tools in this section to classify a site. The Beckingham et al. (1996) field guide should be consulted to ensure the characteristics outlined in the field guide suitably match the interpreted information for the ecosite phase.

Graphic keys and tables that summarize existing resource information sources applicable to ecological classification are presented in Figures 7–10 and Tables 5–7. The codes used in the keys and tables are found in the Glossary or Appendixes 1–7. The overview key helps the user select the appropriate key and table leads for further ecological classification. These groups were originally derived from forest inventory systems (Lindenas 1985) and ultimately modified and used in the field guide (Beckingham et al. 1996). For continuity, they are used in this document to make the keys less overwhelming and more user-friendly. Ecosystems can be homogeneous or heterogeneous and, in either case, can be entirely deciduous in composition (hardwood), entirely coniferous in composition (softwood), a mixture of deciduous and coniferous (mixedwood) or non-forested. To prevent duplication of information, deciduous and mixedwood keys are presented together.

The graphic keys use a variety of information sources to assist in the identification of ecosite phases. Ecological systems are dynamic and variable across landscape areas; therefore differentiating ecosite phases using specific characteristics and their magnitudes is difficult because various magnitudes of several factors interact to create the current ecosite phase. Hence, the keys must be used as a general guide and not relied upon exclusively. The keys consist of decision boxes that provide typical vegetation composition (i.e., canopy), dominant landform, soil characteristics, site characteristics, and where applicable, the most frequent topographic position for ecosite phases or groups of ecosite phases. The decision process directs the user to the proper ecosite phase by describing general ecological conditions that help differentiate ecosite phases.

When two or more features are listed in the decision boxes, they are not weighted or ranked in any particular order. The user must decide whether one, some, most, or all of the conditions must be met in order to make the decision. This is because high values for some factors can compensate for low values for other factors. For example, any of topographic position, water table depth, slope, aspect or soil texture may be dominant in influencing the ecological function of a particular site. Which factor or combination of factors are most important will become more clear as the available information is synthesized and compiled. The user is encouraged to integrate their own experiences and interpretations into the classification process. The user must be able to move freely between decision boxes in all directions until the appropriate ecosite phase is determined. Remember to compare the final key selection (i.e., ecosite phase) with its associated fact sheet from the field guide and check ecosite phases that are adjacent on the edatope (Beckingham et al. 1996). If the site does not match, then trace your steps within the key and try another lead. As well as using the graphic key, consult the appropriate information table (Tables 5, 6, and 7). These tables provide a direct way to compare ecosite phase differences for ecologically similar groups across ecological gradients and among topographic position, landform, soil classification, soil association and dominant soil textures.

In some cases, the graphic keys result in multiple ecosite phases. This occurs when particular ecological features of the site cannot be assessed from aerial photographs and resource maps (e.g., ground cover of shrubs or forbs). This information must be obtained from ground truthing; however, by pre-classifying areas using the methods outlined in this manual the user can classify or tag areas by the amount of ground work required, thus, focusing resources and saving time and money. This system was designed to complement field work and not replace it.

Both the Saskatchewan forest inventory (Lindenas 1985) and the ecosite classification (Beckingham et al. 1996) systems use tree canopy and soil characteristics to define the classification groups. The forest inventory system allows users to link a particular stand type to general soil textures and soil drainage patterns; however, soil drainage and texture are mapped over larger areas than single stands. Inconsistencies arise when mapped information describes the prevailing or dominant conditions in the polygon but does not match the specific conditions determined at a field sample site. For example, Thorpe (1990) found that only 32% of the forest inventory polygons sampled had the same mapped drainage class as determined in the field. The ecosite classification places similar stand types in groups where common soil and site attributes are shared (ecosite). It is a field-based taxonomic system that uses site-specific soil, site, and vegetation information to classify an area. While the forest inventory is an aerial photo interpretation derived mapping system and the ecosite classification system is a field-based taxonomic system, a natural overlap between the systems occurs (Beckingham et al. 1996). The correlation between the systems, however, is far from perfect (Beckingham et al. 1996; Silviba Services Ltd. and Dendron Resource Surveys Inc. 1996).

An attempt was made in the spring of 1997 to correlate forest inventory units with the ecosite classification system. This involved representatives from Geographic Dynamics Corp., Saskfor MacMillan Limited Partnership, and the Department of Environment and Resource Management. Results from this correlation process are presented in Appendix 8. Forest inventory map units were assigned the most probable ecosite. In some cases, an ecosite complex was used when forest inventory units grossly overlapped two or more ecosite types. The correlation is a gross tool for assigning forest inventory map units to ecosites using a "best-fit" comparison. Assigning forest inventory units to ecosites does not transform the forest inventory into an ecosite classification system because it is still based largely on an interpretation of aerial photographs and retains the limitations of this classification methodology. The value and limitations of both systems are important when interpreting sites and applying models generated from this correlation. Thus, we recommend that forest inventory maintenance maps are used in conjunction with soil maps, aerial photographs, and the field guide to interpret the correct ecosite phase from resource material with the potential to further investigate and verify the classification in the field.

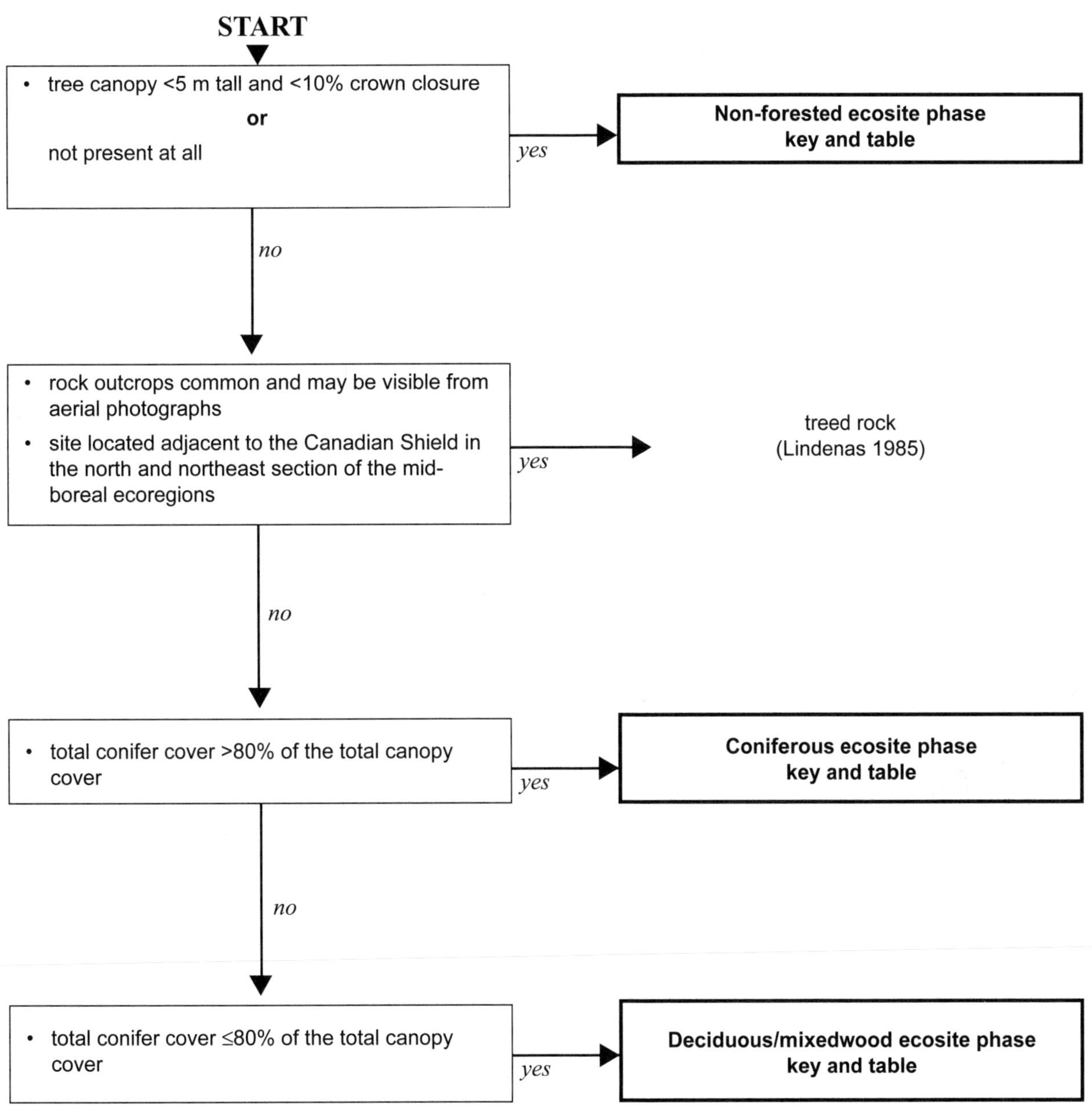

Note: Tamarack is a deciduous conifer tree. When it occurs as a tree, it is found in the coniferous key and table and when it occurs as a shrub or stunted tree, it is found in the non-forested key and table.

Figure 7. Overview key to non-forested, coniferous, and deciduous/mixedwood ecosite phase keys and tables.

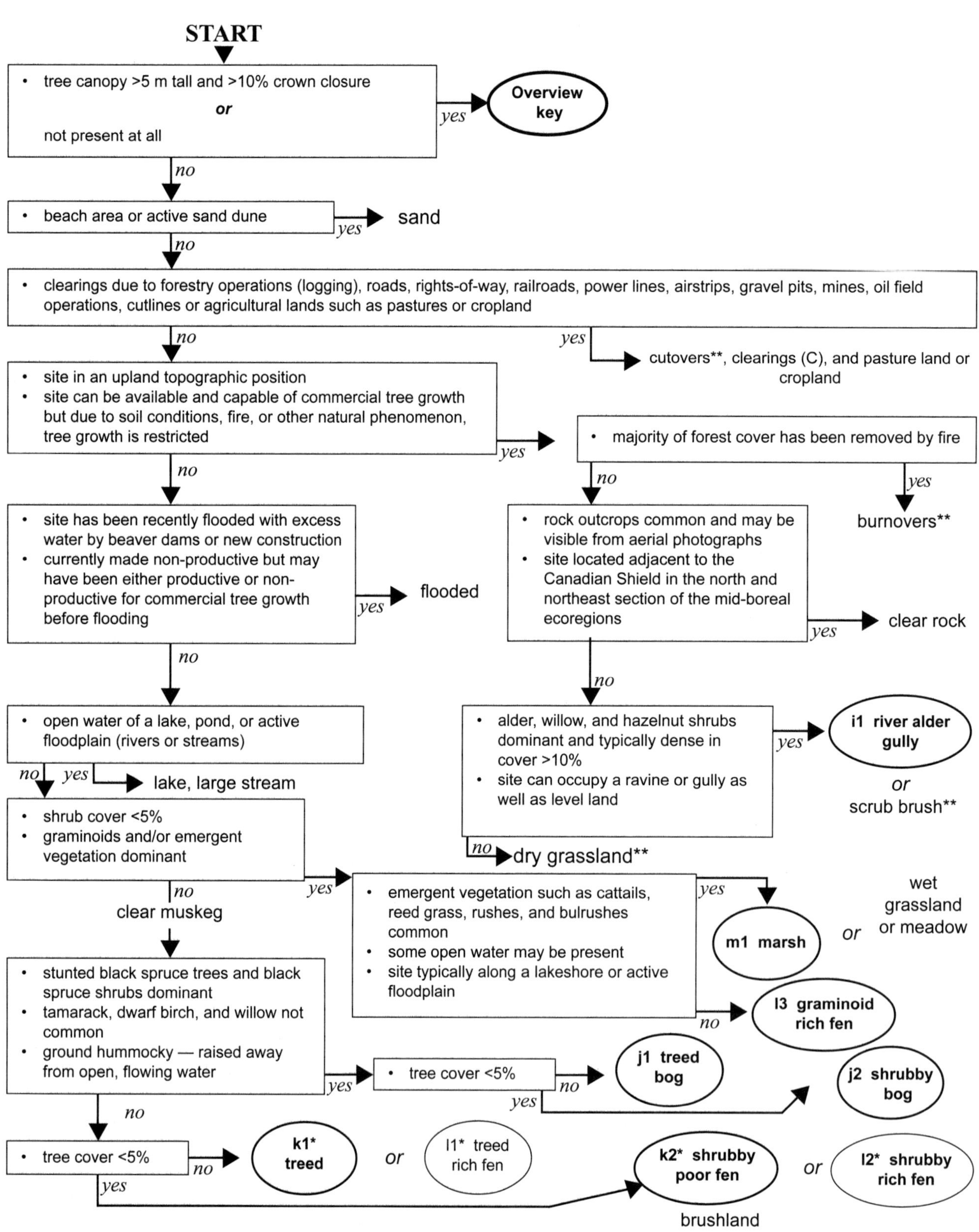
START
tree canopy >5 m tall and >10% crown closure
or
not present at all
yes
Overview key
no
beach area or active sand dune
yes
sand
no
clearings due to forestry operations (logging), roads, rights-of-way, railroads, power lines, airstrips, gravel pits, mines, oil field operations, cutlines or agricultural lands such as pastures or cropland
no
yes
cutovers**, clearings (C), and pasture land or cropland
site in an upland topographic position
site can be available and capable of commercial tree growth but due to soil conditions, fire, or other natural phenomenon, tree growth is restricted
yes
majority of forest cover has been removed by fire
no
yes
burnovers**
no
site has been recently flooded with excess water by beaver dams or new construction
currently made non-productive but may have been either productive or non-productive for commercial tree growth before flooding
yes
flooded
rock outcrops common and may be visible from aerial photographs
site located adjacent to the Canadian Shield in the north and northeast section of the mid-boreal ecoregions
yes
clear rock
no
no
open water of a lake, pond, or active floodplain (rivers or streams)
alder, willow, and hazelnut shrubs dominant and typically dense in cover >10%
site can occupy a ravine or gully as well as level land
yes
i1 river alder gully
or
scrub brush**
no
yes
lake, large stream
shrub cover <5%
graminoids and/or emergent vegetation dominant
no
dry grassland**
no
clear muskeg
yes
emergent vegetation such as cattails, reed grass, rushes, and bulrushes common
some open water may be present
site typically along a lakeshore or active floodplain
yes
m1 marsh
or
wet grassland or meadow
no
l3 graminoid rich fen
stunted black spruce trees and black spruce shrubs dominant
tamarack, dwarf birch, and willow not common
ground hummocky — raised away from open, flowing water
yes
tree cover <5%
no
j1 treed bog
yes
j2 shrubby bog
no
tree cover <5%
no
k1* treed
or
l1* treed rich fen
yes
k2* shrubby poor fen
or
l2* shrubby rich fen
brushland

* requires ground truthing to distinguish from similar ecosite phases.

** interpreter can classify the site to ecosite level only with physical site characteristics such as soil texture and drainage. If previous knowledge of canopy structure exists, then this information can also be useful in determining ecosite.

Figure 8. **Key to the non-forested ecosite phases.** Land types other than those distinguished in Beckingham et al. (1996) are adapted from Lindenas (1985).

Table 5. Summary information on the non-forested ecosite phases

<table>
<tr><th>Ecosite phase</th><th>Moisture regime</th><th>Nutrient regime</th><th>Drainage</th><th>Topographic position</th><th>Parent material/ landform</th><th>Common great groups</th><th>Common soil associations</th><th>Effective texture</th></tr>
<tr><td>i1</td><td>hygric to subhydric</td><td>rich to very rich</td><td>poor</td><td>depression, toe, and level</td><td>fluvial, swamp</td><td>Humic Regosol, Gleysol, (Fibrisol), (Humisol)</td><td>Alluvium, Stream Fen</td><td>variable</td></tr>
<tr><td>j2</td><td rowspan="4">hygric to hydric</td><td>very poor to poor</td><td rowspan="5">poor to very poor</td><td rowspan="5">depression and level</td><td rowspan="4">organic</td><td>Fibrisol, Mesisol, (Organic Cryosol)</td><td rowspan="4">Flat Bog, Bowl Bog, Horizontal Fen, Bowl Fen, Stream Fen</td><td rowspan="4">organic</td></tr>
<tr><td>k2</td><td>poor to medium</td><td>Fibrisol, Mesisol, (Static Cryosol)</td></tr>
<tr><td>l2</td><td rowspan="2">medium to very rich</td><td rowspan="2">Fibrisol, Mesisol, (Humic Gleysol), (Gleysol)</td></tr>
<tr><td>l3</td></tr>
<tr><td>m1</td><td>subhydric to hydric</td><td>very rich</td><td>fluvial, lacustrine, moraine, organic</td><td>Gleysol, (Fibrisol)</td><td>Marsh, Stream Fen, Alluvium</td><td>organic and variable mineral</td></tr>
</table>

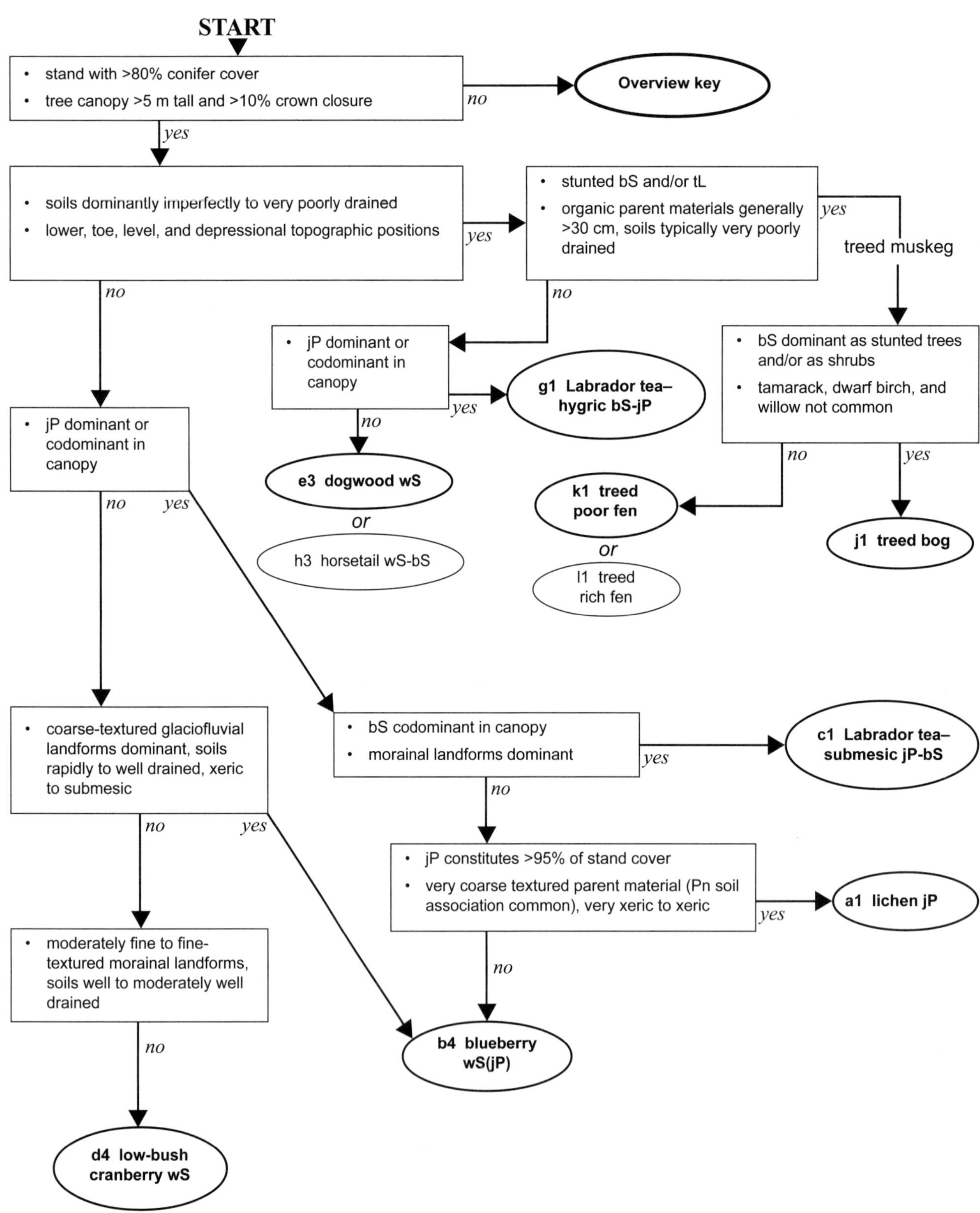

Figure 9. Key to the coniferous ecosite phases. Note: e3 dogwood wS, h3 horsetail wS–bS, k1 treed poor fen and l1 treed rich fen all require ground truthing to distinguish them from similar ecosite phases.

Table 6. Summary information on the coniferous ecosite phases

Ecosite phase	Tree species	Moisture regime	Nutrient regime	Drainage	Topographic position	Parent material/ landform	Common great groups	Common soil associations	Effective texture
a1	jP	very xeric to subxeric	very poor to poor	very rapidly to rapidly	crest to midslope and level	glaciofluvial, eolian	Eutric Brunisol	Pine, Kakwa	coarse
b4	wS, jP, (tA)	xeric to submesic	poor to medium	rapidly to well		moraine, glaciofluvial	Eutric Brunisol, Gray Luvisol		coarse to moderately coarse
c1	jP, bS, (wB)	submesic to subhygric	very poor to medium	rapidly to mod. well	upper to midslope and level		Gray Luvisol, Eutric Brunisol	Loon River, Bittern lake, Waterhen, Dorintosh, Waitville, Piprell	variable
d4	wS, (bF), (tA), (wB), (bP)		medium	well to mod. well	upper to lower slope and level	moraine, glaciolacustrine			moderately fine to fine
e3	wS, bF, (wB), (tA)	mesic to hygric	medium to rich	mod. well to imperfectly	mid to lower slope, level, and depression		Gray Luvisol, Humic Gleysol, (Mesisol)	La Corne, Bittern Lake, Alluvium, Arbow, Stream Fen, Horizontal Fen	
g1	bS, jP, (wS), (tA), bP	subhygric to subhydric	very poor to medium	imperfectly to poorly	lower slope, level, depression, and toe	moraine, glaciofluvial, glaciolacustrine	Luvic Gleysol, Gleysol, Gray Luvisol	Loon River, Bittern Lake, La Corne, Dorintosh, Bow River, Alluvium, Arbow	variable
h3	wS, bS, (tL), (bP)		medium to rich			glaciolacustrine	Humic Gleysol, Gleysol, (Humisol)	Alluvium, Arbow, Bowl Fen	
j1	bS, (tL)	hygric to hydric	very poor to poor	poorly to very poorly		organic	Mesisol, Fibrisol, (Luvic Gleysol)	Bowl Bog, Stream Bog, Flat Bog, Horizontal Fen, Stream Fen, Bowl Fen, Arbow	organic
k1	tL, bS		poor to medium				Mesisol, Fibrisol, (Humic Gleysol)		
l1	tL, (bS)		medium to very rich				Mesisol, Fibrisol, (Gleysol)		

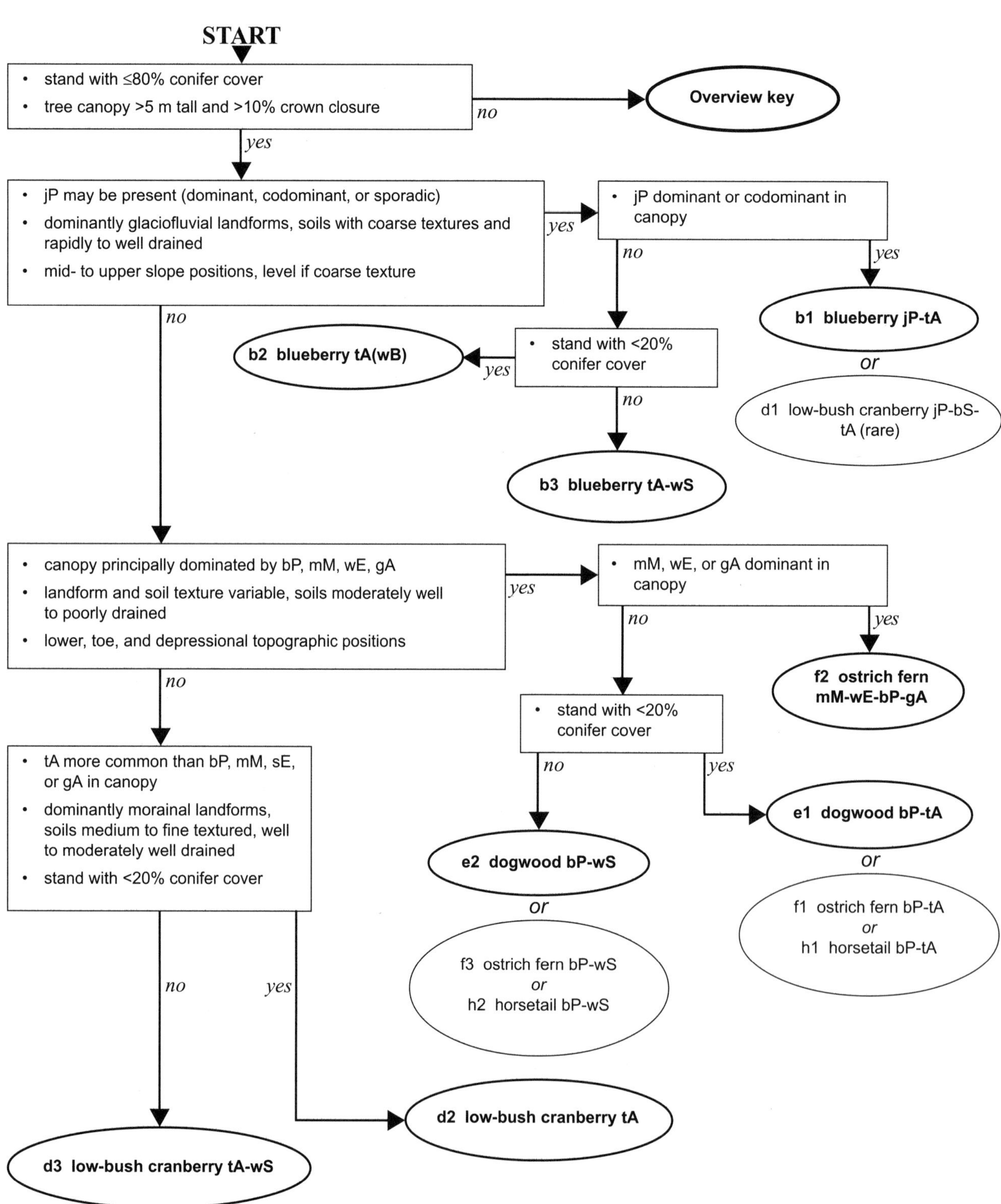

Figure 10. Key to the deciduous and mixedwood ecosite phases. Note: b1 blueberry jP-tA, d1 low-bush cranberry jP-bS-TA, e2 dogwood bP-wS, f3 ostrich fern bP-wS, h2 horsetail bP-wS, e1 dogwood bP-tA, f1 ostrich fern bP-tA, and h1 horsetail bP-tA all require ground truthing to distinguish them from similar ecosite phases.

Table 7. Summary information on the deciduous and mixedwood ecosite phases

<table>
<tr><th>Ecosite phase</th><th>Tree species</th><th>Moisture regime</th><th>Nutrient regime</th><th>Drainage</th><th>Topographic position</th><th>Parent material/ landform</th><th>Common great groups</th><th>Common soil associations</th><th>Effective texture</th></tr>
<tr><td>b1</td><td>jP, tA, (wS), (bS), (wB)</td><td rowspan="3">subxeric to submesic</td><td rowspan="3">poor to medium</td><td rowspan="3">rapidly to well</td><td rowspan="3">crest to midslope and level</td><td rowspan="3">glaciofluvial, fluvial, moraine</td><td rowspan="3">Eutric Brunisol, Dystric Brunisol, Gray Luvisol</td><td>Loon River, Waterhen, Bittern Lake, Kewanoke</td><td>variable</td></tr>
<tr><td>b2</td><td>tA, wB, (wS), (jP)</td><td>Waterhen, Loon River, Bodmin, Pine, Kewanoke, Hillwash, Beach</td><td rowspan="2">coarse to moderately coarse</td></tr>
<tr><td>b3</td><td>tA, wS, (jP), (bP)</td><td>Bodmin, Bittern Lake, Pine</td></tr>
<tr><td>d1</td><td>jP, (tA), (wS), (bF), (bS)</td><td rowspan="3">submesic to subhygric</td><td rowspan="3">medium</td><td rowspan="3">well to mod. well</td><td rowspan="3">upper to midslope and level</td><td rowspan="2">moraine, glaciofluvial</td><td rowspan="3">Gray Luvisol, Eutric Brunisol</td><td>Loon River, Bittern Lake</td><td>variable</td></tr>
<tr><td>d2</td><td>tA, bP, wB, wS</td><td>Bittern Lake, Waitville, Loon River, Pine</td><td rowspan="2">moderately fine to fine</td></tr>
<tr><td>d3</td><td>wS, tA, wB, bF, bS, bP</td><td>moraine, glaciolacustrine</td><td>Loon River, Bittern Lake, Dorintosh, Waitville</td></tr>
<tr><td>e1</td><td>bP, tA, wB, wS</td><td rowspan="2">mesic to hygric</td><td rowspan="2">medium to rich</td><td rowspan="5">mod. well to imperfectly</td><td rowspan="7">lower slope, depression, level, and toe</td><td>moraine, glaciolacustrine, fluvial</td><td rowspan="2">Luvic Gleysol, Gray Luvisol, Regosol</td><td>Dorintosh, Waitville, Waterhen, Loon River, Kakwa, Alluvium, Arbow</td><td rowspan="2">moderately fine to fine</td></tr>
<tr><td>e2</td><td>wS, bP, tA bF, wB</td><td>moraine, lacustrine, glaciolacustrine</td><td>Loon River, Waitville, Alluvium, Beach</td></tr>
<tr><td>f1</td><td>bP, tA, wB</td><td rowspan="5">subhygric to hygric</td><td rowspan="3">rich to very rich</td><td>glaciolacustrine, moraine, fluvial</td><td rowspan="2">Humic Gleysol, Luvic Gleysol, Regosol, Gray Luvisol, Gleysol, Humic Regosol</td><td>Loon River, Arbow</td><td rowspan="3">variable</td></tr>
<tr><td>f2</td><td>mM, wE, bP, (wB), (gA)</td><td rowspan="2">fluvial</td><td>Sipanok, Meadow</td></tr>
<tr><td>f3</td><td>bP, wS, bF, wB, tA</td><td rowspan="3">Gleysol, Humic Gleysol, Humic Regosol, (Eutric Brunisol), (Mesisol), (Gray Luvisol)</td><td rowspan="2">Alluvium</td></tr>
<tr><td>h1</td><td>tA, bP, (wS)</td><td rowspan="2">medium to rich</td><td rowspan="2">mod. well to poorly</td><td>fluvial, colluvium</td><td>moderately fine to fine</td></tr>
<tr><td>h2</td><td>wS, bP, bS, tA, (bF)</td><td>fluvial, lacustrine, glaciolacustrine</td><td>Alluvium, La Corne, Bowl Fen</td><td>variable</td></tr>
</table>

6.0 CLASSIFICATION EXAMPLES OF ECOSITE PHASES IN TYPICAL TOPOSEQUENCES

This section provides five examples of typical toposequences found in the mid-boreal ecoregions of Saskatchewan. An attempt was made to include as many different ecosite phases as possible; however, some units naturally appear less frequently within the landscape and may not be presented. The examples are derived from actual aerial photographs and try to show a variety of moisture and nutrient situations. Note that in the Mid-Boreal Lowland Ecoregion, the river alder gully (i1) ecosite phase is not restricted to ravines or gullies as it is in the Mid-Boreal Upland Ecoregion.

The user is encouraged to follow the steps outlined in Section 4.0 and the keys and tables in Section 5.0 to successfully determine ecosite phase.

Each example includes:

- an aerial photo stereogram of the selected area and example transect;
- forest inventory and soil survey maps of the area; and
- interpreted landscape profile and summary tables.

The user can see how the information is interpreted from the primary sources (aerial photographs, forest inventory, soil survey, and field guide). They can also develop an understanding of the relationships and the ecological importance of the variables used to identify ecosite phase from aerial photographs.

Plant silhouettes and parent material patterns used in the landscape profiles are outlined in Figure 11 while common codes and symbols from map sources can be found in Table 8. Other information found within the examples can be found in the Glossary or Appendixes 1–8.

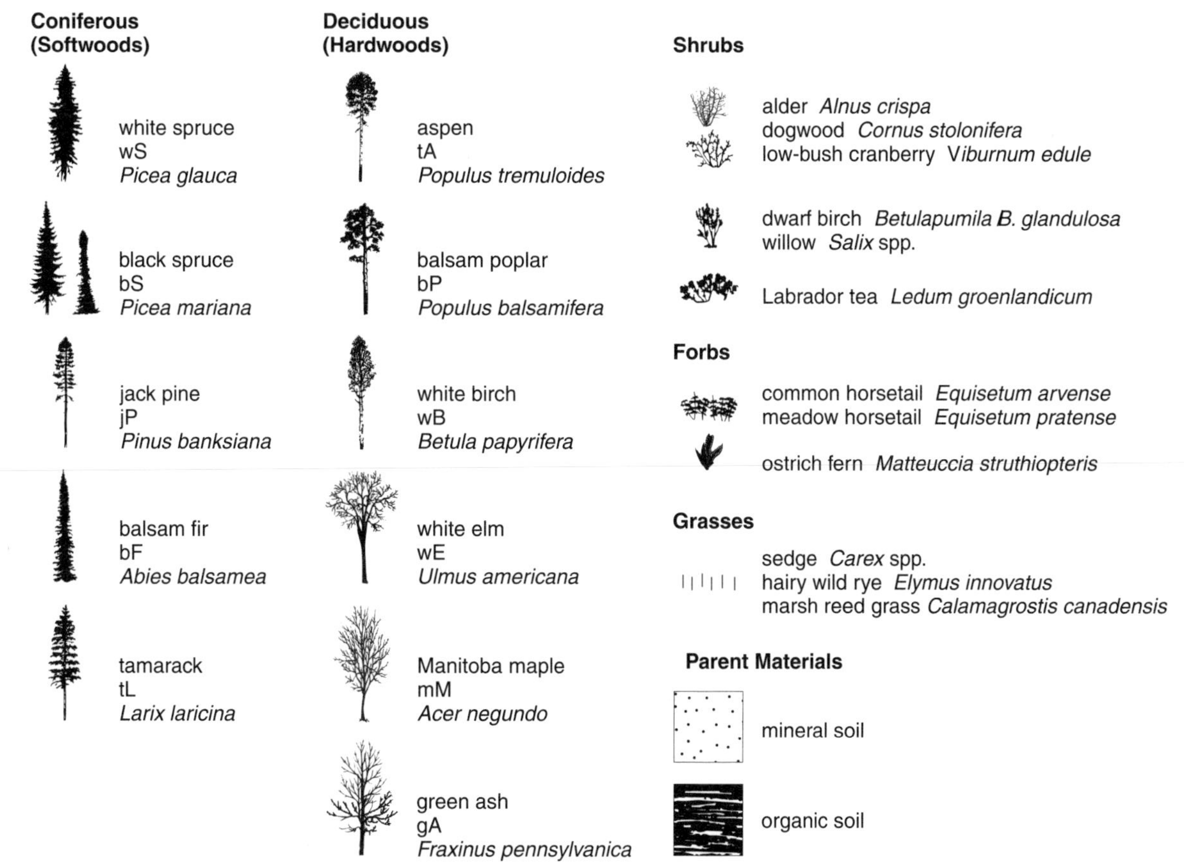

Figure 11. Plant silhouettes and parent material patterns.

Table 8. Common codes and symbols from forest inventory maintenance maps and soil survey maps

General Species Association	
S	Softwood (>75% softwoods by volume)
H	Hardwood (< 25% softwoods by volume)
SH	Mixedwood (50–75% softwoods by volume)
HS	Mixedwood (25–50% softwoods by volume)

Crown Closure		Height	
10–29%	A	2.5–7.5 m	5
30–55%	B	7.5–12.5 m	10
55–80%	C	12.5–17.5 m	15
≥ 80%	D	17.5–22.5 m	20
		≥ 22.5 m	25

Common Map Symbols			
Treed muskeg		Scrub brush	SB
Clear muskeg		Clearing	C
Brushland		Road	
Meadow		Water	
Forest cover type/Soil boundary		River, stream, or creek	
Drainage boundary		Transect	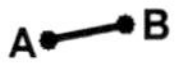 A—B
Cutover	 C.O.		

Drainage		Texture	
Very rapidly drained	VR	Coarse	C
Rapidly drained	R	Moderately coarse	MC
Well drained	W	Moderately fine	MF
Moderately well drained	MW	Fine	F
Imperfectly drained	I		
Poorly drained	P		

Stone Classes		Slope Classes and Topography	
Stone-free	St0	Very gently sloping or gently undulating	0.5–2% slope
Slightly stony land	St1	Gently sloping or undulating	2–5% slope
Moderately stony land	St2	Moderately sloping or gently rolling	6–9% slope
Very stony land	St3	Steeply sloping or strongly rolling	10–15% slope
Exceedingly stony land	St4	Steeply sloping or strongly rolling	16–30% slope
Excessively stony land	St5	Very steeply sloping or hilly	30–60% slope

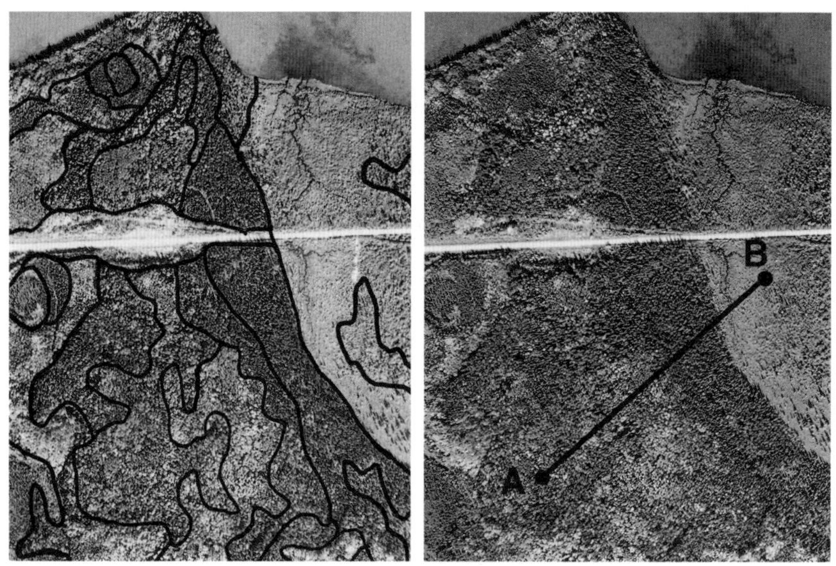

Aerial photo stereogram (1:12 500)
Imaged September 12, 1987; film: Kodak Double-X Aerographic 2405; camera: Zeiss RMK A15; focal length: 153.647 mm.

Forest inventory maintenance map (1:12 500)

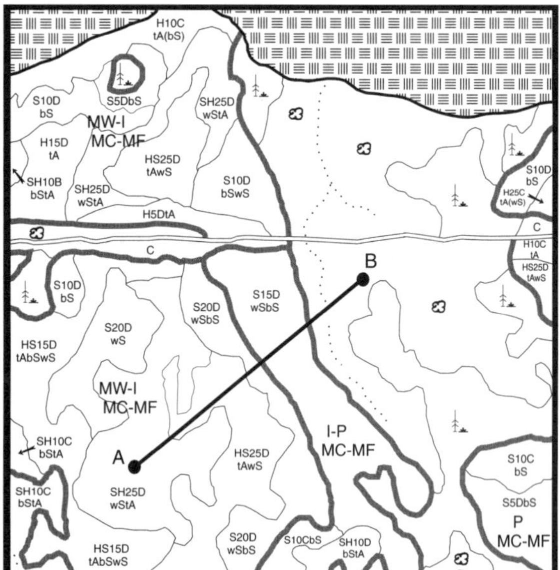

Soil survey map (1:84 480)

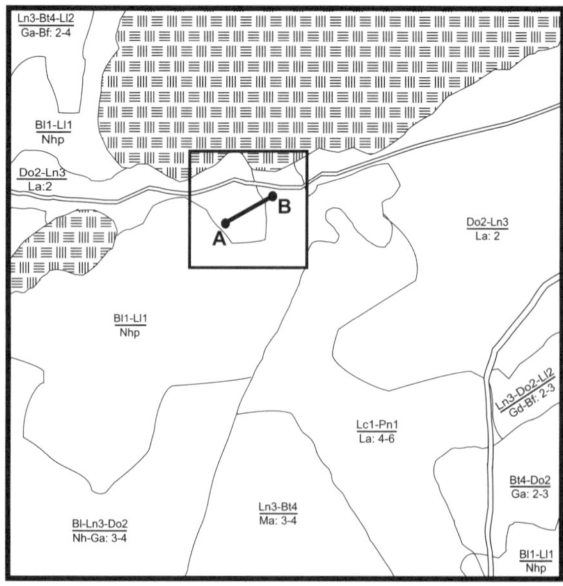

Figure 12. A toposequence of the Mirasty Lake area including an aerial photo stereogram, a forest inventory maintenance map, a soil survey map, and ecosite attributes.

Figure 12 continued.

Ecosite attributes

FIELD GUIDE Ecosite and soil type classification[a]					
Ecosite Phase	d3 low-bush cranberry tA-wS		d4 low-bush cranberry wS	e3 dogwood wS	k2 shrubby poor fen
Soil Type	SM4 Moist/Fine Loamy-Clayey				SR Organic
LANDSCAPE PROFILE					
	A				B
FOREST INVENTORY					
Species Association	**SH25D** tA wS	**HS25D** tA wS	**S20D** wS bS	**S15D** wS bS	**Brushland**
General Designation	SH = mixedwood 50% ≤ all SWD <75%; 25% ≤ all HWD <50%	HS = mixedwood 25% ≤all SWD <50% 50% ≤all HWD <75%	S = softwood 75% ≤all SWD ≤100% 0% ≤ all HWD <25%		not applicable
Specific Designation	wS > any other SWD tA > any other HWD		wS and bS > any other SWD by vol. % wS > % bS; bS >25%		not applicable
Height Class	22.5 m < HC25		17.5 m < HC20 ≤ 22.5 m	12.5 m < HC15 ≤17.5 m	not applicable
Crown Closure	80% < Class D ≤100%				not applicable
Drainage and Texture Label	**MW–I** **MC–MF** moderately well to imperfectly drained moderately coarse to moderately fine			**I–P** **MC–MF** imperfectly to poorly drained moderately coarse to moderately fine	organic soils
SOIL SURVEY[b]					
	Do2–Ln3 **La:2**				**Bl1–Ll1** **Nhp**
Map Units	**Do2** = Dorintosh Dominant Soil Subgroup = O.GL Significant Soil Subgroups = Gleysolic subgroups – peaty phase Parent Material = Moderately fine to fine-textured GL deposits. **Ln3** = Loon River Dominant Soil Subgroup = O.GL Significant Soil Subgroups = Gleyed Gray Luvisols, Gleysols – peaty phase Parent Material = Medium to moderately fine textured, weakly to moderately calcareous glacial till				**Bl1** = Bagwa Lake Dominant Soil Subgroup = TY.M Parent Material = Fen peat **Ll1** = Lavallee Lake Dominant Soil Subgroup = TY.M Significant Soil Subgroup = TY.F Parent Material = Forest peat frequently overlain by shallow sphagnum peat and sometimes underlain by fen peat.
Landform	**La** = GL plains; knoll and depression				**Nhp** = Mixed horizontal and patterned fen
Slope Class	**2** = 0.5–2%, gently undulating				not applicable

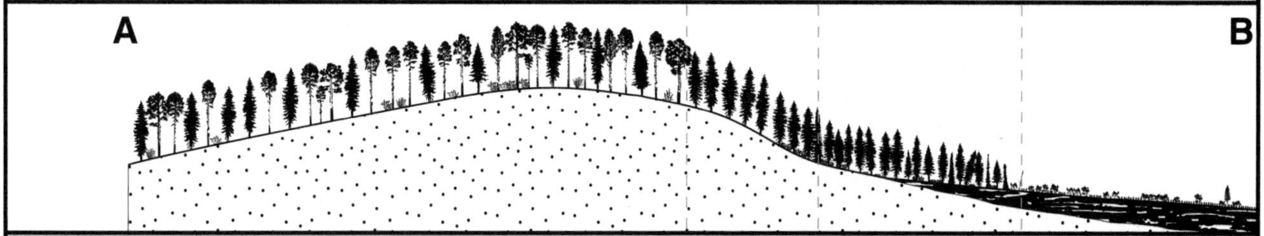

[a] Beckingham et al. 1996.

[b] Padbury 1984.

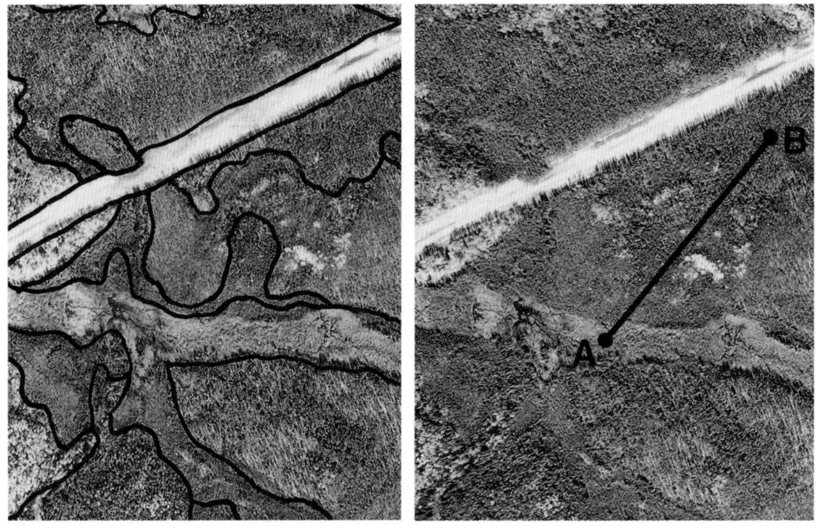

Aerial photo stereogram (1:12 500)
Imaged September 17, 1984; film: Kodak Double-X Aerographic 2405;
camera: Wild RC-8; focal length: 152.034 mm.

Forest inventory maintenance map
(1:12 500)

H15D
tA
H15D
tA
H15D
tA
HS20D
tAwS
S15C
jP
HS15C
tAbS
W-MW
C-MC
C
B
S20C
jP
S15B
jP
H20C
tA
C
H20CtA
S10D
bS
S15D
jPbS
W-MW
C-MC
S15D
jPbS
A
S10C
jPbS
W-MW
MC-MF
HS20D
tAwSjP
W-MW
MC-MF
S10D
bS
S15C
jPbS
S10D
jPbS
S15B
jPbS
S10D

Soil survey map
(1:84 480)

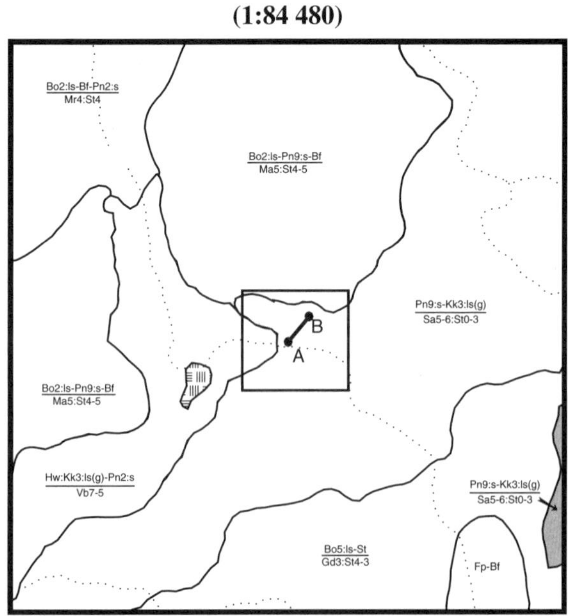

Figure 13. A toposequence in the area south of Wapawekka Lake including an aerial photo stereogram, a forest inventory maintenance map, a soil survey map, and ecosite attributes.

Figure 12 continued.

Ecosite attributes

FIELD GUIDE Ecosite and soil type classification[a]				
Ecosite Phase	k2 shrubby poor fen	g1 Labrador tea–hygric bS-jP	c1 Labrador tea–submesic jP-bS	a1 lichen jP
Soil Type	SR Organic	SWp Wet/Peaty	SD1 Dry/Sandy	SV1 Very Dry/Sandy
LANDSCAPE PROFILE				
A				B
FOREST INVENTORY				
Species Association	**Brushland**	**S10D bS**	**S15D jP bS**	**S20C jP**
General Designation	not applicable	S = softwood 75% ≤ all SWD ≤100% 0% ≤ all HWD <25%		
Specific Designation	not applicable	% bS > any other SWD by vol. % jP <25%	% jP and % bS > any other SWD by vol.; % jP > % bS; % bS >25%	% jP > any other SWD by vol.; % bS <25%
Height Class	not applicable	7.5 m < HC10 ≤12.5 m	12.5 m < HC15 ≤17.5 m	17.5 m < HC20 ≤22.5 m
Crown Closure	not applicable	80% < Class D ≤100%		55% < Class C ≤80%
Drainage and Texture Label	organic soils	**W–MW** **C–MC** well to moderately well drained coarse to moderately coarse textured		
SOIL SURVEY[b]				
	Pn9: s–Kk3: ls(g) **Sa5–6: St0–3**			
Map Units	**Pn9** = Pine Dominant Soil Subgroups = E.EB, O.R Significant Soil Subgroups = GLE.EB, GL.R, Gleysolic soils – peaty phase Parent Material = Coarse-textured, weakly to noncalcareous, GF and GL sands, some of which have been reworked by the wind Texture = s = sand **Kk3** = Kewanoke Dominant Soil Subgroup = E.EB Parent Material = Coarse-textured, weakly to noncalcareous, gravelly GF deposits Texture = ls(g) = loamy sand (gravelly)			
Landform	**Sa** = Fluvial–Lacustrine Plains, knoll and depression			
Slope Class	**5–6** = 10–30%, strongly to steeply sloping or moderately to strongly rolling			
Stone Class	**St0–3** = Stone free to very stony			

[a] Beckingham et al. Spec. Rep. 6.

[b] Head et al. 1981.

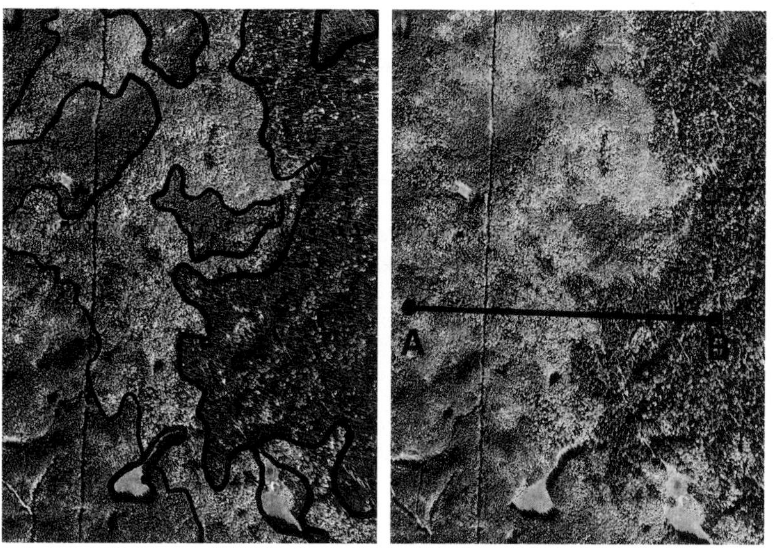

Aerial photo stereogram (1:12 500)
Imaged September 15, 1984; film: Kodak Double-X Aerographic 2405; camera: Wild RC-8; focal length: 152.034 mm.

Forest inventory maintenance map (1:12 500)

Soil survey map (1:84 480)

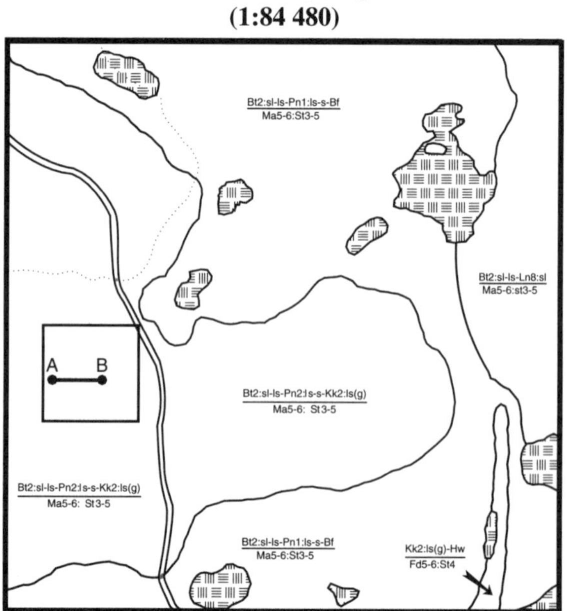

Figure 14. A toposequence in the area west of Montreal Lake including an aerial photo stereogram, a forest inventory maintenance map, a soil survey map, and ecosite attributes.

Figure 14 continued.

Ecosite attributes

FIELD GUIDE Ecosite and soil type classification[a]			
Ecosite Phase	b1 blueberry jP-tA	d2 low-bush cranberry tA	d3 low-bush cranberry tA-wS
Soil Type	SV1 Very Dry/Sandy	SM4 Moist/Fine Loamy-Clayey	
LANDSCAPE PROFILE			
FOREST INVENTORY			
Species Association	**SH15D** **jP tA**	**H15D** **tA**	**SH25C** **wS tA**
General Designation	SH = mixedwood 50% ≤ all SWD <75% 25% ≤all HWD <50%	H = hardwood 0% ≤all SWD <25% 75% ≤all HWD ≤100%	SH = mixedwood 50% ≤ all SWD <75% 25% ≤all HWD <50%
Specific Designation	% jP > any other SWD % tA > any other HWD	% tA > any other HWD by vol.	% wS > any other SWD % tA > any other HWD
Height Class	12.5 m < HC15 ≤17.5 m		22.5 m < HC25
Crown Closure	80% < Class D ≤100%		55% < Class C ≤80%
Drainage and Texture Label	**W–MW** **MC–MF** well to moderately well drained moderately coarse to moderately fine textured		
SOIL SURVEY[b]			
	Bt2: sl–ls-Pn2: ls–s-Kk2: ls(g) **Ma5–6: St3–5**		
Map Units	**Bt2** = Bittern Lake Dominant Soil Subgroup = BR.GL Significant Soil Subgroup = O.GL Parent Material = Medium to moderately fine textured, moderately calcareous glacial till, overlain by moderately coarse to medium-textured materials. Texture = sandy loam (ls) to loamy sand (ls) **Pn2** = Pine Dominant Soil Subgroups = E.EB, O.R Parent Material = Coarse-textured, weakly to noncalcareous, GF and GL sands, some of which have been reworked by the wind Texture = loamy sand (ls) to sand (s) **Kk2** = Kewanoke Dominant Soil Subgroup = E.EB Significant Soil Subgroup = O.EB Parent Material = Coarse-textured, weakly to noncalcareous, gravelly GF deposits Texture = ls(g) = loamy sand (gravelly)		
Landform	**Ma** = Moraine, knob-and-kettle		
Slope Class	**5–6** = 10–30%, strongly to steeply sloping or moderately to strongly rolling		
Stone Class	**St3–5** = very stony to excessively stony		

[a] Beckingham et al. 1996.

[b] Head et al. 1981.

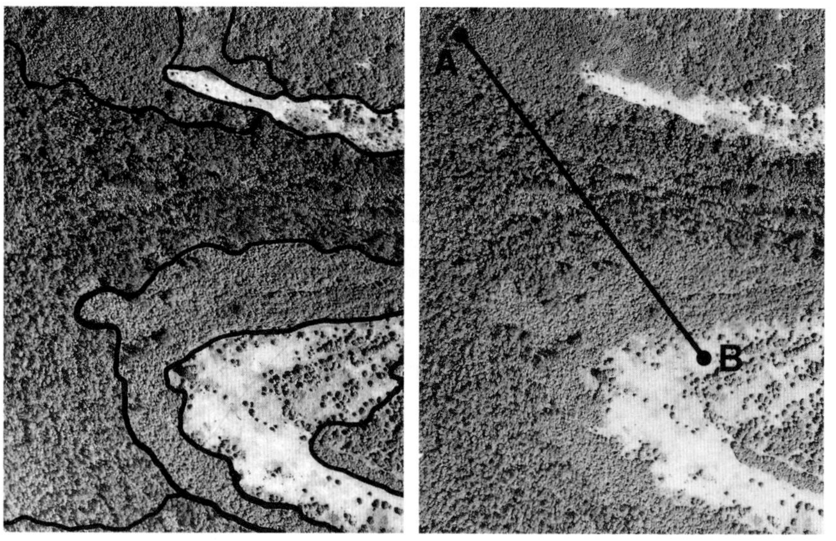

Aerial photo stereogram (1:12 500)
Imaged June 6, 1977; film: Kodak Double-X Aerographic 2405;
camera: Wild RC-10; focal length: 153.08 mm.

Forest inventory maintenance map
(1:12 500)

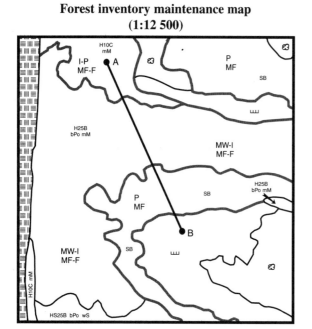

H10C
mM
I-P
MF-F
A
P
MF
SB
H25B
bPo mM
MW-I
MF-F
P
MF
SB
H25B
bPo mM
B
MW-I
MF-F
SB
H10C mM
HS25B bPo wS

Soil survey map
(1:84 480)

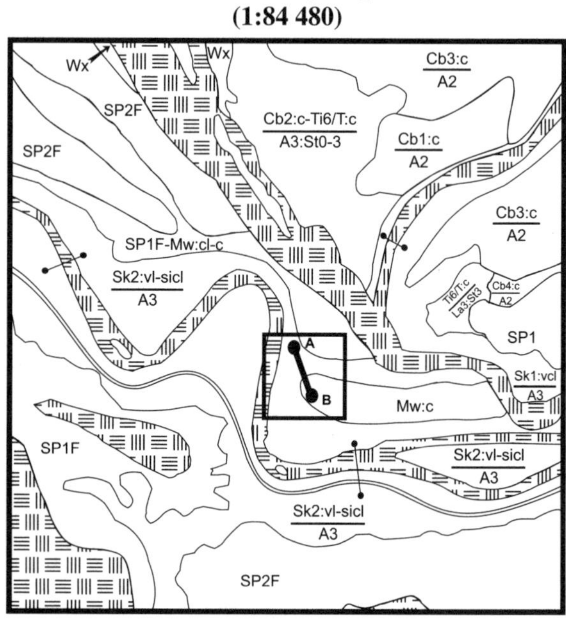

Figure 15. A toposequence between Old Channel and Saskatchewan rivers including an aerial photo stereogram, a forest inventory maintenance map, a soil survey map, and ecosite attributes.

Figure 15 continued.

Ecosite attributes

Field Guide	**Ecosite and soil type classification**[a]			
Ecosite Phase	f2 ostrich fern mM-wE-bP-gA		i1 river alder gully	l3 graminoid rich fen
Soil Type	SM4 Moist/Fine Loamy-Clayey		SWm Wet/Mineral	SR Organic
Landscape Profile	 A			B
Forest Inventory				
Species Association	**H10C mM**	**H25B bPo mM**	**Scrub (SB)**	**Meadow**
General Designation	H = hardwood 0% ≤ all SWD <25% 75% ≤all HWD ≤100%		not applicable	not applicable
Specific Designation	% mM > any other HWD by vol.	% bPo > any other HWD by vol. % mM >10% but <25% of vol. of stand	not applicable	not applicable
Height Class	7.5 m < HC10 ≤12.5 m	22.5 m < HC25	not applicable	not applicable
Crown Closure	55% < Class C ≤80%	30% < Class B ≤55%	not applicable	not applicable
Drainage and Texture Label	**I–P MF–F** imperfectly to poorly drained moderately fine to fine textured	**MW–I MF–F** moderately well to imperfectly drained moderately fine to fine textured	**P MF** poorly drained moderately fine textured	organic soils
Soil Survey[b]				
	Sk2: vl–sicl A3		**Mw: c**	
Map Units	**Sk2** = Sipanok Dominant Soil Subgroup = R.DG Significant Soil Subgroups = CU.R, GLCU.R Parent Material = Medium to fine-textured, strongly calcareous, stratified, recent levee deposits Texture = very fine sandy loam (vl) and silty clay loam (sicl)		**Mw** = Meadow Dominant Soil Subgroups = Gleysolic soils Parent Material = Variable textured, glacial and recent deposits associated with local intermittently flooded areas Texture = clay (c)	
Landform	**A** = Alluvial Plains, unpatterned alluvial plain		not applicable	
Slope Class	**3** = 2–5%, gently sloping to gently and roughly undulating		not applicable	

[a] Beckingham et al. 1996.

[b] Ayers et al. 1978.

Aerial photo stereogram (1:12 500)
Imaged July 2, 1981; film: Kodak Infrared Aerographic 2424; camera: Wild RC-10; focal length: 153.02 mm.

Forest inventory maintenance map (1:12 500)

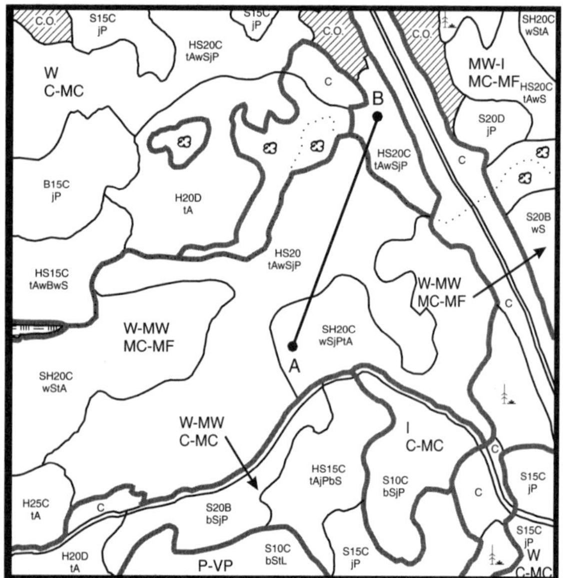

Soil survey map (1:84 480)

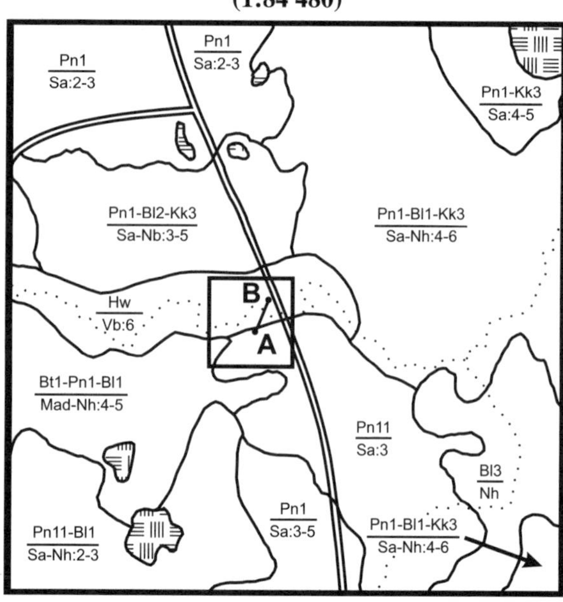

Figure 16. A toposequence west of Highway 903 at Broad Creek including an aerial photo stereogram, a forest inventory maintenance map, a soil survey map, and ecosite attributes.

Figure 16 continued.

Ecosite attributes

FIELD GUIDE Ecosite and soil type classification[a]					
Ecosite Phase	b3 blueberry tA-wS	d3 low-bush cranberry tA-wS	e2 dogwood bP-wS	*open water*	e2 dogwood bP-wS
Soil Type	SV1 Very Dry/Sandy	SD2 Dry/Coarse Loamy	SWm Wet/Mineral		SM2 Moist/Coarse Loamy

LANDSCAPE PROFILE

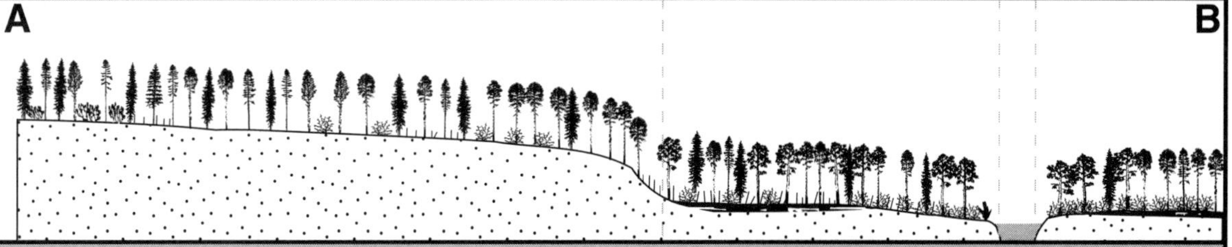

A B

FOREST INVENTORY				
Species Association	**SH20C** **wS jP tA**	**HS20C** **tA wS jP**	*streams or rivers*	**HS20C** **tA wS jP**
General Designation	SH = mixedwood 50% ≤all SWD <75% 25% ≤ all HWD <50%	HS = mixedwood 25% ≤all SWD <50% 50% ≤all HWD <75%		
Specific Designation	wS > any other SWD jP >10% but <25% tA > any other HWD	tA > any other HWD wS > any other SWD % jP > 10% but < 25%		
Height Class	17.5 m < HC20 ≤22.5 m			
Crown Closure	55% < Class C ≤80%			
Drainage and Texture Label	**M–MW** **C–MC** well to moderately well drained coarse to moderately coarse			

SOIL SURVEY[b]				
	Pn11 **Sa:3**	**Hw** **Vb:6**		
Map Units	**Pn11** = Pine Dominant Soil Subgroup = E.EB Parent Material = Coarse-textured, sandy glaciofluvial and glaciolacustrine deposits	**Hw** = Hillwash Dominant Soil Subgroups = Luvisolic, Brunisolic, and organic soils Parent Material = Variable-textured deposits associated with valleys and escarpments		
Landform	**Sa** = Knoll and depression	**Vb** = Valley or glacial spillway with eroded slopes		
Slope Class	**3** = 2–5%, roughly undulating	6 = 16–30%, strongly rolling		

[a] Beckingham et al. 1996.

[b] Padbury et al. 1984.

7.0 REFERENCES

Agriculture Canada Expert Committee on Soil Survey. 1987. The Canadian system of soil classification. 2nd ed. Agric. Can., Ottawa, Ontario. Publ. 1646.

Alberta Forest Service. 1963. A guide to photographic interpretation. Edmonton, Alberta.

Alberta Forestry, Lands and Wildlife. 1991. Ecological land survey site description manual. Resource Information Branch, Land Information and Services Division, Forestry, Lands and Wildlife, Edmonton, Alberta.

Avery, T.E. 1962. Interpretation of aerial photographs: an introductory college textbook and self-instruction manual. Burgess Publishing Co., Minneapolis, Minnesota.

Avery, T.E. 1978. Forester's guide to aerial photo interpretation. U.S. Dep. Agric. For. Serv., Washington, D.C. Agriculture Handb. 308.

Ayres, K.W.; Anderson, D.W.; Ellis, J.G. 1978. The soils of the northern provincial forest in the Pasquia Hills and Saskatchewan portion of The Pas Map Area (63-E and 63-F Sask.). Sask. Inst. Pedol., Saskatoon, Saskatchewan. Publ. SF4.

Beckingham, J.D.; Nielsen, D.G.; Futoransky, V.A. 1996. Field guide to ecosites of the mid-boreal ecoregions of Saskatchewan. Nat. Resour. Can., Can. For. Serv., Northwest Reg., North. For. Cent., Edmonton, Alberta. Spec. Rep. 6.

Carleton, T.J.; Maycock, P.F. 1981. Understorey–canopy affinities in boreal forest vegetation. Can. J. Bot. 59:1709–1716.

Daubenmire, R. 1968. Soil moisture in relation to vegetation distribution in the mountains of northern Idaho. Ecology 49:431–438.

Daubenmire, R. 1973. A comparison of approaches to the mapping of forest land for intensive management. For. Chron. 49:87–92.

Daubenmire, R. 1976. The use of vegetation in assessing the productivity of forest lands. Bot. Rev. 42(2):115–143.

Dix, R.L.; Swan, J.M.A. 1971. The roles of disturbance and succession in upland forest at Candle Lake. Can. J. Bot. 49:657–676.

Ellis, R.A. 1986. Understory development in aspen-white spruce forests in northern Alberta. M.Sc. thesis. Department of Botany. Univ. Alberta, Edmonton, Alberta.

Farrar, J.L. 1995. Trees in Canada. Fitzhenry and Whiteside Ltd., Markham, Ontario and Nat. Resour. Can., Can. For. Serv., Ottawa, Ontario.

Gimbarzevsky, P. 1973. Analysis and interpretation of aerial photographs in the evaluation of biophysical environment. Div. Cont. Ed., Univ. Calgary, Calgary, Alberta.

Hall, F.C. 1978. Applicability of rangeland management concepts to forest-range in the Pacific Northwest. Pages 496–499 *in* D.N. Hyder, Ed. Proc. First Int. Rangeland Congr., August 14–18, 1978, Denver, Colorado. Soc. Range Manage., Denver, Colorado.

Hedrick, D.W. 1975. Grazing mixed conifer clearcuts in northeastern Oregon. Rangeman's J. 2:6–9.

Heller, R.C.; Doverspike, G.E.; Aldrich, R.C. 1964. Identification of tree species on large-scale panchromatic and color aerial photographs. U.S. Dep. Agric., For. Serv., Washington, D.C. Agric. Handb. 261.

Howard, J.A. 1970. Aerial photo-ecology. Faber and Faber, London.

Jeglum, J.K; Boissonneau, A.N. 1977. Air photo interpretation of wetlands, northern clay section, Ontario. Can. Dep. Environ., Can. For. Serv., Great Lakes For. Res. Cent. Rep. 0-X-269.

Johnson, D.; Kershaw, L; MacKinnon, A.; Pojar, J., Eds. 1995. Plants of the western boreal forest and aspen parkland. Lone Pine Publ., Edmonton, Alberta.

Lacate, D.S., compiler. 1969. Guidelines for biophysical land classification. Can. Dep. Fish. For., For. Serv., Ottawa, Ontario. Publ. 1264.

Lee, S.C. 1924. Factors controlling forest successions at Lake Itasca, Minnesota. Bot. Gaz. 78:129–174.

Lindenas, D.G. 1985. Specifications for the interpretation and mapping of aerial photographs in the forest inventory section. Saskatchewan Parks and Renew. Resourc. For. Div., Prince Albert, Saskatchewan. Intern. Doc.

Lutz, H.J.; Caporaso, A.P. 1959. Vegetation and topographic situation as indicators of forested land classes in air-photo interpretation of the Alaska interior. U.S. Dep. Agric., For. Serv., Jeaneau, Alaska. Stn. Pap. 10.

Martin, N.D. 1959. An analysis of forest succession in Algonquin Park, Ontario. Ecol. Monogr. 29:187–218.

Moss, E.H. 1953. Forest communities in northwestern Alberta. Can. J. Bot. 31:212–252.

Mueller-Dombois, D. 1964. The forest habitat types in southeastern Manitoba and their application to forest management. Can. J. Bot. 42:1417–1444.

Padbury, G.A.; Acton, D.F. 1994. Ecoregions of Saskatchewan 1:2,000,000 scale map. Saskatchewan Property Manage. Corp., Cent. Surv. Mapping Agency, Regina, Saskatchewan.

Pase, C.P. 1958. Herbage production and composition under immature ponderosa pine stands in the Black Hills. J. Range Manage. 11:238–243.

Pojar, J.; Meidinger, D.; Klinka, K. 1991. Concepts. Pages 9–37 *in* Meidinger, D. and J. Pojar, Eds. Ecosystems of British Columbia. British Columbia Minist. For., Victoria, British Columbia. Spec. Rep. Ser. 6.

Raup, H.M.; Denny, C.S. 1950. Photo interpretation of the terrain along the southern part of the Alaska Highway. U.S. Geol. Surv. Bul. 963-D. Gov. Print. Office, Washington, D.C.

Rowe, J.S. 1956. Uses of undergrowth plant species in forestry. Ecology 37(3):461–472.

Sayn-Wittgenstein, L. 1960. Recognition of tree species on air photographs by crown characteristics. Can. Dep. For., For. Res. Div., Ottawa, Ontario. Tech. Note 95.

Sayn-Wittgenstein, L. 1978. Recognition of tree species on aerial photographs. Can. Dep. Environ., Can. For. Serv., For. Manage. Inst., Ottawa, Ontario. Inf. Rep. FMR-X-118.

Silviba Services Ltd. and Dendron Resource Surveys Inc. 1996. A survey of undisturbed lands within the Prince Albert Model Forest to test the Draft 2 Field guide to ecosites of the mid-boreal upland ecoregion of Saskatchewan. Silviba Services Ltd., Prince Albert, Saskatchewan and Dendron Resource Surveys Inc., Ottawa, Ontario. Prepared for Prince Albert Model Forest Association Inc., May 24, 1996. Unpubl. rep.

Stoeckeler, E.G. 1952. Trees of interior Alaska; their significance as soil and permafrost indicators. Corps Eng., U.S. Army, St. Paul, Minnesota.

Strong, W.L.; Leggat, K.R. 1991. Ecoregions of Alberta. Prepared for Alberta Forestry, Lands and Wildlife, Land Information Services Division by Ecological Land Surveys Ltd. Edmonton, Alberta. Unpubl. rep.

Thorpe, J. 1990. An assessment of Saskatchewan's system of forest site classification. Saskatchewan Res. Counc. Publ. E-2530-1-E-90.

Uresk, D. W.; Severson, K.E. 1989. Understory-overstory relationships in ponderosa pine forests, Black Hills, South Dakota. J. Range Manage. 42(3):203–208.

van Kesteren, A.R. 1992. Air photo interpretation of Damman forest types on calcareous terrain in western Newfoundland. For. Can., Nfld. Lab. Reg., St. John's, Newfoundland. Inf. Rep. N-X-286.

Viro, P.J. 1961. Evolution of site fertility. Unasylva 15:91–97.

Alluvium: Fluvial sediment deposits found in flooded areas or stream channels.

Aspect: The orientation of a slope as determined by the points of a compass. Aspect, in combination with slope, is important in predicting the amount of solar radiation a site receives. A level site has no aspect.

Brushland: Non-productive land with greater than 50% of the vegetation cover composed of large shrubs (willow, alder, and/or dwarf birch).

Calcareous: The presence of calcium carbonate ($CaCO_3$) as indicated by effervescence (bubbling) when soil material is treated with dilute hydrochloric acid (HCl).

Clearing: Non-productive land where tree cover amounts to less than 10% crown closure. These are typically areas of anthropogenic activities including roads, railroads, air strips, or gravel pits.

Coarse Fragments: The volumetric percentage of rock portions larger than 2 mm in diameter within a soil matrix. Generally they are grouped into three diameter classes: gravels, cobbles, and stones.

Codominant Plant Species: One of several species that contribute the greatest cover in a plant community type.

Colluvium (C): A mixture of weathered soil and/or geological materials transported downslope by gravitational forces and deposited at the base of a slope.

Coniferous: Cone-bearing evergreen trees or shrubs. Note: Tamarack is a deciduous conifer.

Cover: *See* percent cover.

Deciduous: Trees or shrubs that shed their leaves annually.

Dominant Plant Species: The species that contributes the greatest cover in a plant community.

Drainage: The rapidity and extent of water removal from the soil in relation to additions, especially by surface runoff and by percolation downwards through soil.

Ecoregion: A geographic area that has a distinctive, mature ecosystem on reference sites plus specified edaphic variations as a result of a given regional climate.

Ecosite: Ecological units that develop under similar environmental influences. Ecosites, in the classification system, are groups of one or more ecosite phases that occur within the same portion of the edatope, thus, they are functional units defined by moisture and nutrient regime.

Ecosite Phase: A subdivision of the ecosite based on the dominant tree species in the canopy. On some sites where a tree canopy is lacking, the tallest structural vegetation layer determines the ecosite phase. Generally, ecosite phases are mappable units.

Edaphic: Pertains to the soil, particularly with respect to its influence on plant growth and other organisms together with climate.

Edatope: Moisture/nutrient grid that displays the potential ranges of relative moisture and nutrient conditions and outlines relationships between each of the ecosites.

Eolian (E): Well-sorted, poorly compacted, medium to fine sand and coarse silt sediment that has been transported and deposited by wind. Syn. aeolian.

Floodplain: Flat alluvial area between stream banks subject to inundation by flood water usually once annually.

Fluvial (F): Moderately well sorted sediments of gravels and sands with some fractions of silt and clay, transported and deposited by streams or rivers.

Fluviolacustrine (FL): Lacustrine deposits that have been partially reworked by fluvial processes.

Forested Productive Land: All productive lands that currently support a stand of living trees that can be expected to produce a merchantable stand within a reasonable length of time.

Glaciofluvial (GF): Stratified outwash transported and deposited by glacial meltwaters that flowed on, within, under, or beyond the glacier.

Glaciolacustrine (GL): Stratified sediments with generally alternating light and dark bands deposited in glacial lakes.

Gleyed Soil: An imperfectly or poorly drained soil with characteristics such as dull, gray colors or mottles indicative of periodic conditions of water saturation.

Great Group: Classes of soils within soil orders that are based on properties that reflect differences in the strengths of dominant processes or a major contribution of a process in addition to the dominant one.

Hardwood: Stand type that is dominated by deciduous trees such as aspen, white birch, or balsam poplar. It correlates with the forest inventory species association code (H) which is <25% coniferous by volume, however, has less than 20% total conifer cover in the canopy according to the *Field guide to ecosites of the mid-boreal ecoregions of Saskatchewan*.

HWD: *See* hardwood.

Knob-and-Kettle Landscape: Terrain with irregular topography from scattered hummocks (knobs) and holes (kettles).

Lacustrine (L): Fine sand, silt, and clay sediments deposited on the lake bed or coarser sands that are deposited along a beach by wave action.

Landform: Relief features of the earth's surface produced mainly by erosional and depositional processes.

Landscape Area: A subdivision of the ecoregion based on differences in physiography, surface expression, and the proportion and distribution of soils and plant community types within an area. A landscape area has a recurring pattern of slope, landform, soils, vegetation, and climate.

Levee: Adjacent floodplain deposits to the main channel created during channel inundation and overbank flooding. This deposition is typically heavier nearest the channel and decreases further away. The result is a sloped levee surface away from the main channel.

Meadow: Non-productive land with less than 50% of the vegetation cover composed of shrubs (willow, alder, dwarf birch, tamarack, black spruce, and Labrador tea). Graminoids such as sedges, reed grasses, and rushes dominate the site along with other hydrophytic vegetation (cattails).

Merchantable Stand: Trees that are suitable for sale (greater than 2.5 m in height and a crown closure of greater than 10%).

MBL: Mid-Boreal Lowland Ecoregion

MBU: Mid-Boreal Upland Ecoregion

Mixedwood: Stand type that is a blend of deciduous and coniferous trees. It correlates with the forest inventory species association codes (SH, HS) which are ≥25% and <75% coniferous by volume, however, has ≥20% and ≤80% total conifer cover in the canopy according to the *Field guide to ecosites of the mid-boreal ecoregions of Saskatchewan*.

Moisture/Nutrient Grid: *See* edatope.

Moisture Regime (MR): Represents the available moisture supply for plant growth on a relative scale, ranging from xeric to hydric. It is assessed through an integration of species composition and primary soil and site characteristics. Syn. hygrotope.

Morainal/Till (M): Sediment generally consisting of well-compacted material that is non-stratified and contains a mixture of sand, silt, and clay that has been transported beneath, beside, within, or in front of a glacier.

MR: *See* moisture regime.

Non-Forested Productive Land: Land that is available for or capable of producing a merchantable stand within a reasonable length of time but is not currently supporting a stand of living trees. Included are the following forest inventory units: cutovers, burnovers, open productive lands, and regenerating species.

Non-Productive Land: Land that is not available for or capable of producing a merchantable stand within a reasonable length of time. Included are the following forest inventory units: treed muskeg, treed rock, clear muskeg, clear rock, brushland, meadow, clearing, sand, flooded, lake or large stream, or pasture or crop land.

NR: *See* nutrient regime.

Nutrient Regime (NR): Measure of essential nutrients that are available for plant growth. The determination of nutrient regime requires the integration of many environmental and biotic parameters. Soil nutrient regime occurs on a relative scale ranging from very poor to very rich. Syn. trophotope.

Organic Soil (O): Material of organic origin, in various stages of decay, that has accumulated over time. Generally, the organic matter originates from sedge or *Sphagnum* peat and is subdivided into bog (B), fen (F), swamp (S), marsh (H), or undifferentiated (O) organic components.

Parent Material: The surficial material from which soils are formed. It has characteristics that have an important effect on the soil forming process.

Pedogenesis: The formation of soils.

Physiography: Pertains to the physical landform characteristics. Syn. geomorphology.

Plant Community Type: A subdivision of the ecosite phase and the lowest taxonomic unit in the classification system. While plant community types of the same ecosite phase share vegetational similarities, they differ in their understory species composition and abundance. These differences may not be mappable from aerial photographs but may be important to wildlife, recreation, and other resource sectors.

Poor Fen: An ecosite that is transitional between the rich fen and bog. A poor fen is intermediate in nutrient regime and is similar floristically to the rich fen and bog.

Primary Site Variables: Those ecologically important data specific to a particular site that can be interpreted directly from the landscape. Several primary site variables put together allow for clarification of secondary site variables.

Productive Land: Land that is available for and capable of supporting merchantable stands of timber within a reasonable length of time.

Rich Fen: A peatland with moderately to well decomposed sedge, grass, and reed peat material formed in eutrophic environments. Mineral-rich waters are at or just above the rich fen surface.

Scrub Brush (SB): Land that is available for and capable of producing merchantable stands of living trees within a reasonable length of time, however, is currently dominated by willow, alder, or hazel shrubs.

Secondary Site Variables: Those ecologically important characteristics that must be interpreted from the combination of some or all of the primary site variables.

Slope: Percentage of vertical rise relative to horizontal distance. Zero degrees as percent slope describes a level site and 45° is equivalent to 100% slope.

Softwood: Stand type that is dominated by coniferous trees such as spruce, fir, pine, or tamarack. It correlates with the forest inventory species association code (S) which is ≥75% coniferous by volume, however, has greater than 80% total conifer cover in the canopy according to the *Field guide to ecosites of the mid-boreal ecoregions of Saskatchewan.*

Soil Association: A group of associated soil series developed from similar parent material and occurring under essentially similar climatic conditions.

Soil Complex: Various combinations of different soils developed from a variety of parent materials, but defined by physiographic characteristics such as occurrence in poorly drained areas, or occurrence on steeply sloping valley sides. When mapped, they represent combinations of two or more map units (soil associations), inseparable due to map scale.

Soil Map Unit: A separation, within a soil association, based on the differences in the kind and amount of types of soil (soil series) present.

Soil Series: A group of soils uniform in profile characteristics and developed from the same parent material.

Soil Texture: The relative proportions of sand, silt, and clay in a mineral soil.

Soil Type: Functional taxonomic units used to stratify soils based on soil moisture regime, effective soil texture, organic matter thickness, and solum depth.

Stand: A complex of living trees greater than 2.5 metres in height and having a crown closure of greater than 10%.

Subgroups: Classes of soils within Great Groups which have particular kinds and arrangements of soil horizons, as well as the properties characteristic of the Great Group.

SWD: *See* softwood.

Till: *See* morainal.

Topographic Position: The relative location of a site on the landscape. Positions range from crest to depression.

Vegetation Strata: Layers of plant growth based on morphology and normal height of all species. These layers are used to describe the appearance of the dominant vegetation components of the ecosite phase.

Water Table: The uppermost level of water in a zone of saturated soil that has a long duration of saturation.

APPENDIX 1

ECOLOGICAL UNITS OF THE FIELD GUIDE TO ECOSITES OF THE MID-BOREAL ECOREGIONS OF SASKATCHEWAN[a]

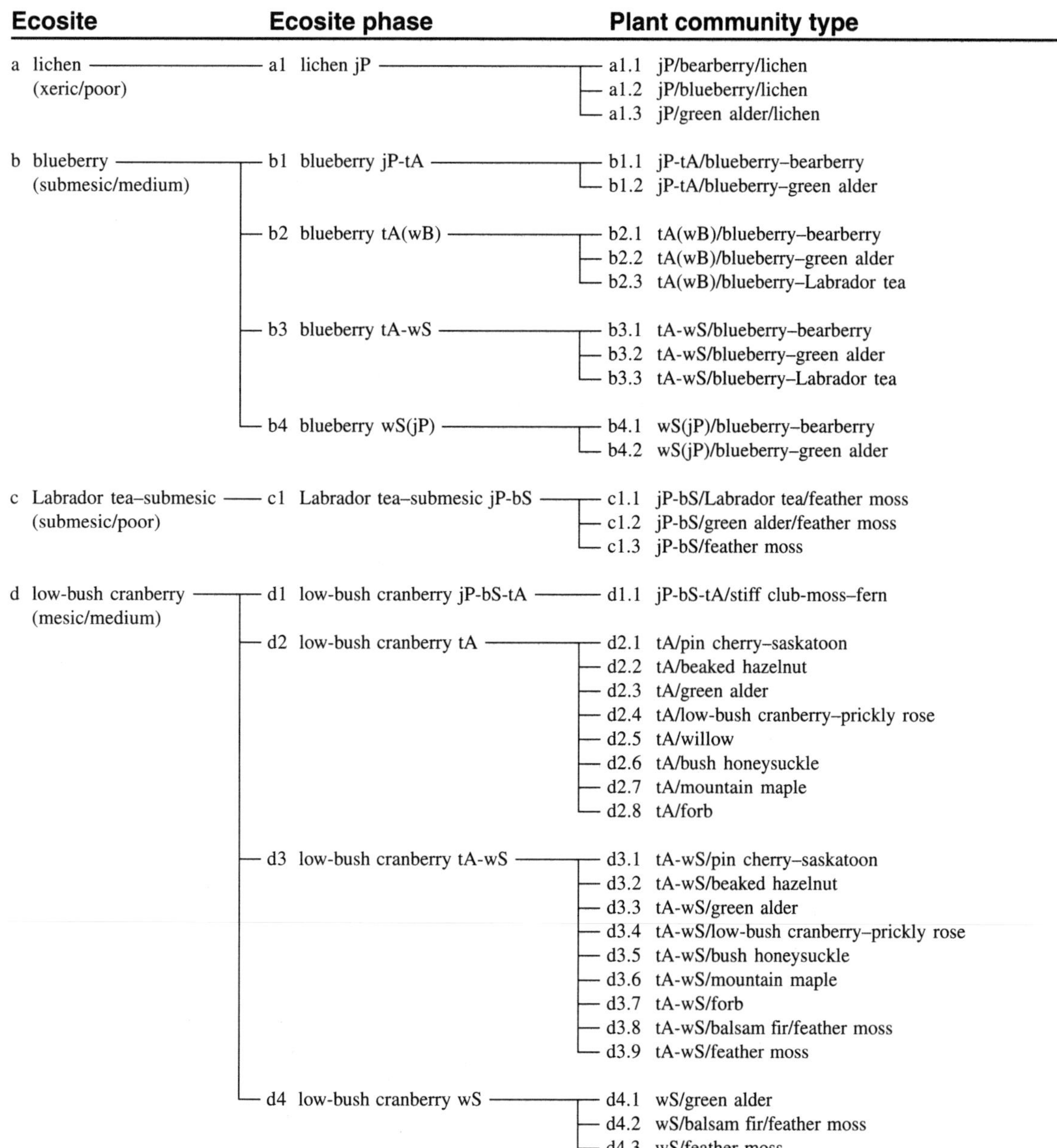

Ecosite	Ecosite phase	Plant community type
a lichen (xeric/poor)	a1 lichen jP	a1.1 jP/bearberry/lichen
		a1.2 jP/blueberry/lichen
		a1.3 jP/green alder/lichen
b blueberry (submesic/medium)	b1 blueberry jP-tA	b1.1 jP-tA/blueberry–bearberry
		b1.2 jP-tA/blueberry–green alder
	b2 blueberry tA(wB)	b2.1 tA(wB)/blueberry–bearberry
		b2.2 tA(wB)/blueberry–green alder
		b2.3 tA(wB)/blueberry–Labrador tea
	b3 blueberry tA-wS	b3.1 tA-wS/blueberry–bearberry
		b3.2 tA-wS/blueberry–green alder
		b3.3 tA-wS/blueberry–Labrador tea
	b4 blueberry wS(jP)	b4.1 wS(jP)/blueberry–bearberry
		b4.2 wS(jP)/blueberry–green alder
c Labrador tea–submesic (submesic/poor)	c1 Labrador tea–submesic jP-bS	c1.1 jP-bS/Labrador tea/feather moss
		c1.2 jP-bS/green alder/feather moss
		c1.3 jP-bS/feather moss
d low-bush cranberry (mesic/medium)	d1 low-bush cranberry jP-bS-tA	d1.1 jP-bS-tA/stiff club-moss–fern
	d2 low-bush cranberry tA	d2.1 tA/pin cherry–saskatoon
		d2.2 tA/beaked hazelnut
		d2.3 tA/green alder
		d2.4 tA/low-bush cranberry–prickly rose
		d2.5 tA/willow
		d2.6 tA/bush honeysuckle
		d2.7 tA/mountain maple
		d2.8 tA/forb
	d3 low-bush cranberry tA-wS	d3.1 tA-wS/pin cherry–saskatoon
		d3.2 tA-wS/beaked hazelnut
		d3.3 tA-wS/green alder
		d3.4 tA-wS/low-bush cranberry–prickly rose
		d3.5 tA-wS/bush honeysuckle
		d3.6 tA-wS/mountain maple
		d3.7 tA-wS/forb
		d3.8 tA-wS/balsam fir/feather moss
		d3.9 tA-wS/feather moss
	d4 low-bush cranberry wS	d4.1 wS/green alder
		d4.2 wS/balsam fir/feather moss
		d4.3 wS/feather moss

[a] Taken from Beckingham, J.D.; Nielsen, D.G.; Futoransky, V.A. 1996. Field guide to ecosites of the mid-boreal ecoregions of Saskatchewan. Nat. Resour. Can., Can. For Serv., Northwest. Reg., North. For. Cent., Edmonton, Alberta. Spec. Rep. 6.

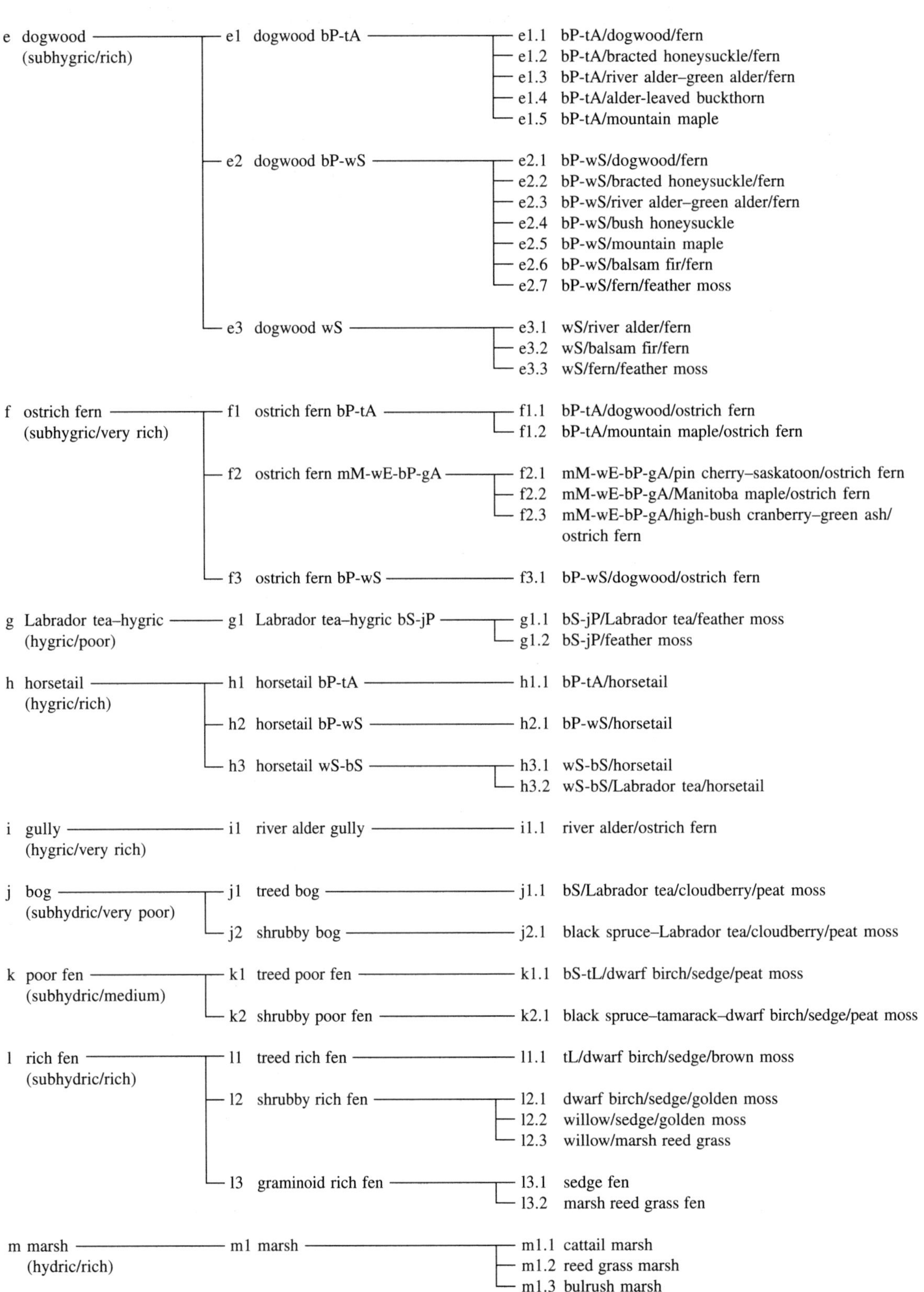

Ecosite	Ecosite phase	Plant community type
e dogwood (subhygric/rich)	e1 dogwood bP-tA	e1.1 bP-tA/dogwood/fern
		e1.2 bP-tA/bracted honeysuckle/fern
		e1.3 bP-tA/river alder–green alder/fern
		e1.4 bP-tA/alder-leaved buckthorn
		e1.5 bP-tA/mountain maple
	e2 dogwood bP-wS	e2.1 bP-wS/dogwood/fern
		e2.2 bP-wS/bracted honeysuckle/fern
		e2.3 bP-wS/river alder–green alder/fern
		e2.4 bP-wS/bush honeysuckle
		e2.5 bP-wS/mountain maple
		e2.6 bP-wS/balsam fir/fern
		e2.7 bP-wS/fern/feather moss
	e3 dogwood wS	e3.1 wS/river alder/fern
		e3.2 wS/balsam fir/fern
		e3.3 wS/fern/feather moss
f ostrich fern (subhygric/very rich)	f1 ostrich fern bP-tA	f1.1 bP-tA/dogwood/ostrich fern
		f1.2 bP-tA/mountain maple/ostrich fern
	f2 ostrich fern mM-wE-bP-gA	f2.1 mM-wE-bP-gA/pin cherry–saskatoon/ostrich fern
		f2.2 mM-wE-bP-gA/Manitoba maple/ostrich fern
		f2.3 mM-wE-bP-gA/high-bush cranberry–green ash/ ostrich fern
	f3 ostrich fern bP-wS	f3.1 bP-wS/dogwood/ostrich fern
g Labrador tea–hygric (hygric/poor)	g1 Labrador tea–hygric bS-jP	g1.1 bS-jP/Labrador tea/feather moss
		g1.2 bS-jP/feather moss
h horsetail (hygric/rich)	h1 horsetail bP-tA	h1.1 bP-tA/horsetail
	h2 horsetail bP-wS	h2.1 bP-wS/horsetail
	h3 horsetail wS-bS	h3.1 wS-bS/horsetail
		h3.2 wS-bS/Labrador tea/horsetail
i gully (hygric/very rich)	i1 river alder gully	i1.1 river alder/ostrich fern
j bog (subhydric/very poor)	j1 treed bog	j1.1 bS/Labrador tea/cloudberry/peat moss
	j2 shrubby bog	j2.1 black spruce–Labrador tea/cloudberry/peat moss
k poor fen (subhydric/medium)	k1 treed poor fen	k1.1 bS-tL/dwarf birch/sedge/peat moss
	k2 shrubby poor fen	k2.1 black spruce–tamarack–dwarf birch/sedge/peat moss
l rich fen (subhydric/rich)	l1 treed rich fen	l1.1 tL/dwarf birch/sedge/brown moss
	l2 shrubby rich fen	l2.1 dwarf birch/sedge/golden moss
		l2.2 willow/sedge/golden moss
		l2.3 willow/marsh reed grass
	l3 graminoid rich fen	l3.1 sedge fen
		l3.2 marsh reed grass fen
m marsh (hydric/rich)	m1 marsh	m1.1 cattail marsh
		m1.2 reed grass marsh
		m1.3 bulrush marsh

EDATOPE (MOISTURE/NUTRIENT GRID) SHOWING THE LOCATIONS OF ECOSITES[a]

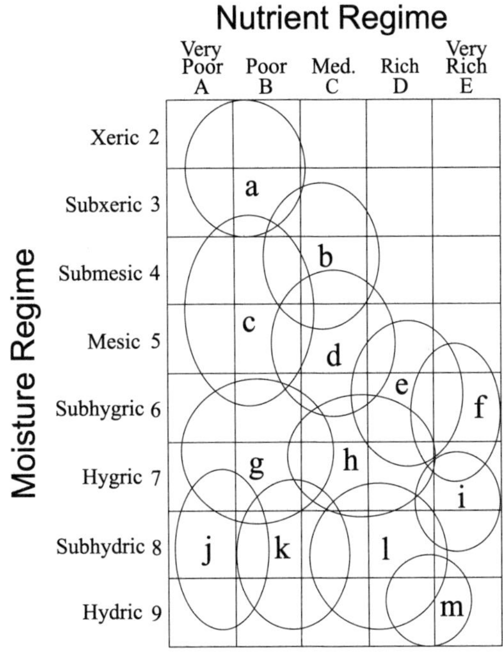

a	lichen	**h**	horsetail
b	blueberry	**i**	gully
c	Labrador tea–submesic	**j**	bog
d	low-bush cranberry	**k**	poor fen
e	dogwood	**l**	rich fen
f	ostrich fern	**m**	marsh
g	Labrador tea–hygric		

[a] Taken from Beckingham, J.D.; Nielsen, D.G.; Futoransky, V.A. 1996. Field guide to ecosites of the mid-boreal ecoregions of Saskatchewan. Nat. Resour. Can., Can. For Serv., Northwest. Reg., North. For. Cent., Edmonton, Alberta. Spec. Rep. 6.

APPENDIX 3

SOIL TYPES OF THE FIELD GUIDE TO ECOSITES OF THE MID-BOREAL ECOREGIONS OF SASKATCHEWAN[a]

SV1	Very Dry/Sandy
SV2	Very Dry/Coarse Loamy
SV3	Very Dry/Silty-Loamy
SV4	Very Dry/Fine Loamy-Clayey
SD1	Dry/Sandy
SD2	Dry/Coarse Loamy
SD3	Dry/Silty-Loamy
SD4	Dry/Fine Loamy-Clayey
SM1	Moist/Sandy
SM2	Moist/Coarse Loamy
SM3	Moist/Silty-Loamy
SM4	Moist/Fine Loamy-Clayey
SMp	Moist/Peaty
SWm	Wet/Mineral
SWp	Wet/Peaty
SR	Organic
SS	Shallow

[a] Taken from Beckingham, J.D.; Nielsen, D.G.; Futoransky, V.A. 1996. Field guide to ecosites of the mid-boreal ecoregions of Saskatchewan. Nat. Resour. Can., Can. For Serv., Northwest. Reg., North. For. Cent., Edmonton, Alberta. Spec. Rep. 6.

APPENDIX 4

TREE SPECIES: ECOLOGY AND IDENTIFICATION

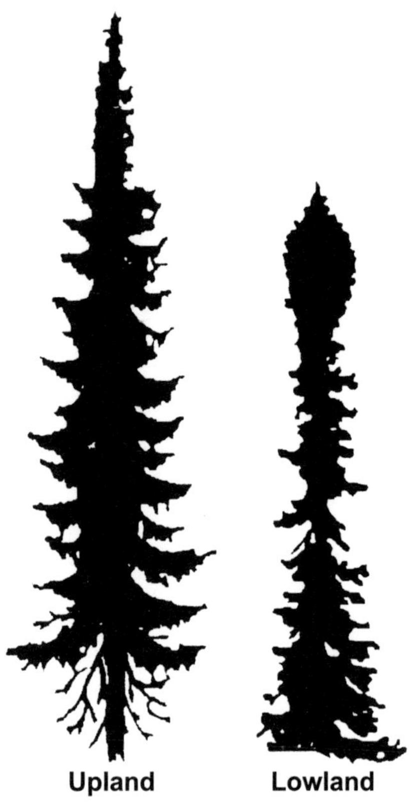

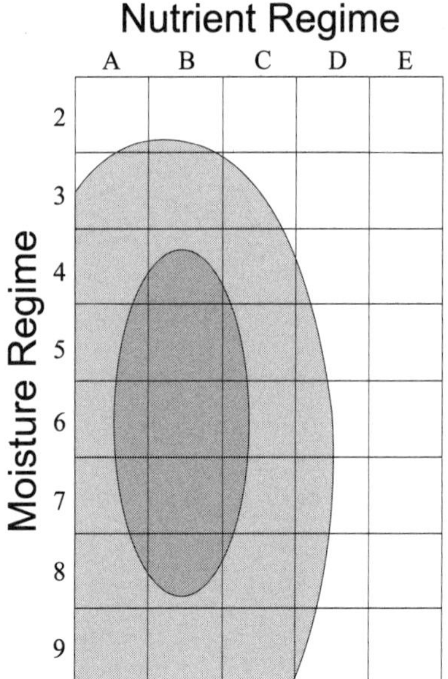

black spruce

bS

Picea mariana (Mill.) B.S.P.

Black spruce (*Picea mariana*) is a coniferous, softwood tree species that is associated with wet, poorly drained soils, and nutrient-poor sites but is also found on upland sites (Farrar 1995; Johnson et al. 1995; Beckingham et al. 1996). On wet sites, such as bogs and fens, black spruce occurs either in pure stands or in association with tamarack. These types of wet sites are found on organic and alluvial landforms, in depressions, and on the fringes of shallow lakes (Gimbarzevsky 1973). On upland sites, black spruce is found in association with jack pine, white spruce, aspen, and balsam fir (Farrar 1995).

Black spruce is narrowly conical to cylindrical in shape (Avery 1962). Dense clumps of foliage and cones at the crown produce a "club top" that may be visible in shadows and on aerial photographs taken at an obtuse angle or on photographs at larger scales. Black spruce stems, when visible, are crooked or inclined usually reflective of the organic material they reside on (Lutz and Caporaso 1959). Black spruce exhibits a dark gray tone on aerial photographs and, although difficult to distinguish, is generally not as dark as white spruce (Heller et al. 1964). Black spruce tips appear as small white dots on aerial photographs and distinct individuals are recognizable. The texture of black spruce on aerial photographs is moderately fine to moderately coarse and dense stands can appear "carpet-like" (Sayn-Wittgenstein 1978).

Black spruce is often difficult to distinguish from white spruce; however, site conditions, species association, and canopy height are clues that provide valuable information.

Black spruce is associated with the following ecosites:

c Labrador tea–submesic

d low-bush cranberry

g Labrador tea–hygric

h horsetail

j bog

k poor fen

white spruce

wS

Picea glauca (Moench) Voss

White spruce (*Picea glauca*) is a coniferous, softwood tree species that grows on dry to moist, well to imperfectly drained soils. White spruce is often found on sites of alluvial, morainal, or lacustrine parent material typically on floodplains, stream banks, and near lakes (Gimbarzevsky 1973). This species can grow in pure stands but is often associated or mixed with aspen, jack pine, black spruce, and balsam fir (Gimbarzevsky 1973; Farrar 1995).

White spruce is conical in shape. Its crown is spire-shaped (Avery 1962), dense and symmetrical with long branches extending almost to the ground (Gimbarzevsky 1973). White spruce casts a triangular shadow with a single, straight trunk and the sharply pointed tips cause the stands to appear moderate to coarse in texture on aerial photographs with individual trees appearing distinct. White spruce displays the highest gray-scale value of the tree species considered here (Heller et al. 1964) and, therefore, exhibits a very dark gray to black tone on aerial photographs.

It is often difficult to distinguish white spruce from black spruce on aerial photographs. Ecological site conditions, tree shape (when visible) and the fact that mature white spruce stands are generally greater in height than those of black spruce (Sayn-Wittgenstein 1978) can be used as clues. Raup and Denny (1950) suggested that white and black spruce can only be distinguished by their topographic position and not on the basis of form. White spruce is also distinguished from balsam fir by having coarser branching, which makes the crown appear more irregular than that of balsam fir (Sayn-Wittgenstein 1978).

White spruce is associated with the following ecosites:

b blueberry

d low-bush cranberry

e dogwood

f ostrich fern

h horsetail

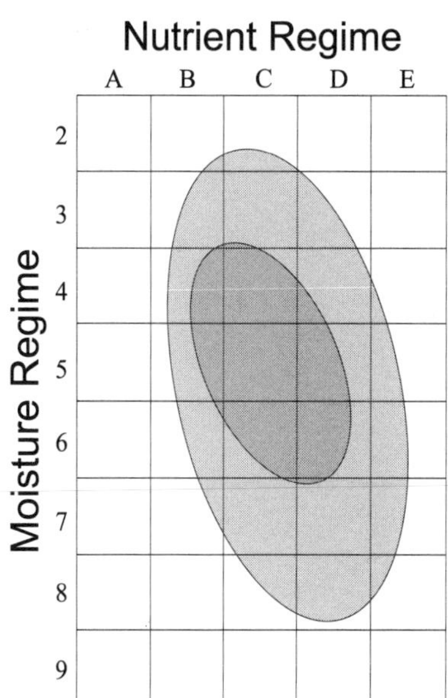

jack pine

jP

Pinus banksiana Lamb.

Jack pine (*Pinus banksiana*) is a coniferous, softwood tree species that is found on well-drained soils and coarse-textured, sandy and gravelly sites (Gimbarzevsky 1973; Farrar 1995; Johnson et al. 1995; Beckingham et al. 1996). These sites have parent materials of glaciofluvial, morainal, lacustrine, and eolian origin (Gimbarzevsky 1973). Jack pine is a fire-origin species and, therefore, is often found in pure, even-aged stands. However, jack pine can also be found in mixed stands with aspen, black spruce, white spruce, balsam fir, balsam poplar, and white birch (Farrar 1995).

Jack pine is irregular, generally conical, and relatively narrow in shape. When viewed from directly above, jack pine is irregular to round in shape with a narrow diameter. On aerial photographs, jack pine appears in moderate gray tones that are significantly lighter than those of white spruce and black spruce. Jack pine stands appear relatively fine-textured and individual trees are distinct with small, narrow, open crowns of many short and twisted branches (Sayn-Wittgenstein 1978). In general, jack pine is readily recognizable.

Jack pine is associated with the following ecosites:

a lichen

b blueberry

c Labrador tea–submesic

d low-bush cranberry

g Labrador tea–hygric

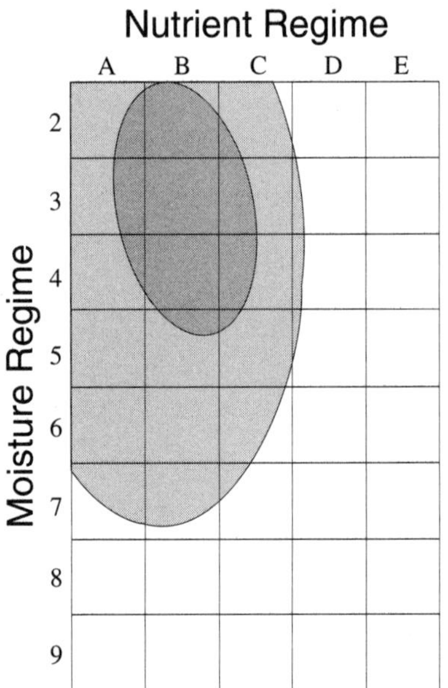

balsam fir

bF

Abies balsamea (L.) Mill.

Balsam fir (*Abies balsamea*) is a coniferous, softwood tree species that grows in mature to over-mature stands on moist morainal and lacustrine landforms (Alberta Forest Service 1963; Gimbarzevsky 1973; Johnson et al. 1995). Balsam fir can occur in pure stands but is commonly mixed with white spruce, black spruce, jack pine, or aspen (Farrar 1995; Johnson et al. 1995).

Balsam fir is conical, tapered, and very symmetrical in shape with a sharply pointed, spire-shaped tip and dense branching (Gimbarzevsky 1973; Avery 1978; Sayn-Wittgenstein 1978). However, its crown has less taper and is slightly more rounded than spruce (Alberta Forest Service 1963). Its foliage is typically very dense. Balsam fir also exhibits a very high gray-scale value appearing very dark to black on aerial photographs (Heller et al. 1964).

Balsam fir is difficult to distinguish from white spruce on aerial photographs. Generally, balsam fir appears more rounded than white spruce when viewed from directly above (Avery 1978). Also, balsam fir is denser and more tapered than white spruce. This can be especially apparent from visible shadows. The tone of balsam fir is similar to that of jack pine, however, separated by a fine, jagged texture and shadows (Gimbarzevsky 1973).

Balsam fir is associated with the following ecosites:

d low-bush cranberry

e dogwood

f ostrich fern

h horsetail

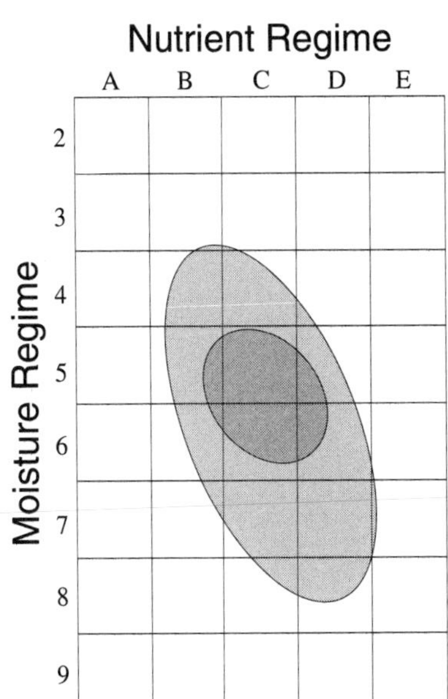

tamarack

tL

Larix laricina (Du Roi) K. Koch

Tamarack (*Larix laricina*) is a deciduous, needle-leaved, softwood tree species. This species is found on wet, poorly drained soils and medium to nutrient-rich sites (Farrar 1995; Beckingham et al. 1996) such as fens. Deep organic deposits, high water table, and wet mineral soils are some of the site characteristics that might be associated with tamarack. Tamarack grows in open areas because it is shade intolerant, as pure stands, or in association with black spruce. In forested conditions, tamarack is dominant with its crown above the general canopy level (Sayn-Wittgenstein 1978).

Tamarack is conical and narrow to oval in shape with very fine foliage. From above it appears broad and irregular in shape. Tamarack appears light to moderate gray in tone and its texture is fine and fuzzy due to its transparent crown (Gimbarzevsky 1973; Avery 1978; Sayn-Wittgenstein 1978). Often, tamarack trunks and shadows of large trees are clearly visible through the lighter toned crown. Tamarack appears quite distinctive on aerial photographs and generally is readily recognizable.

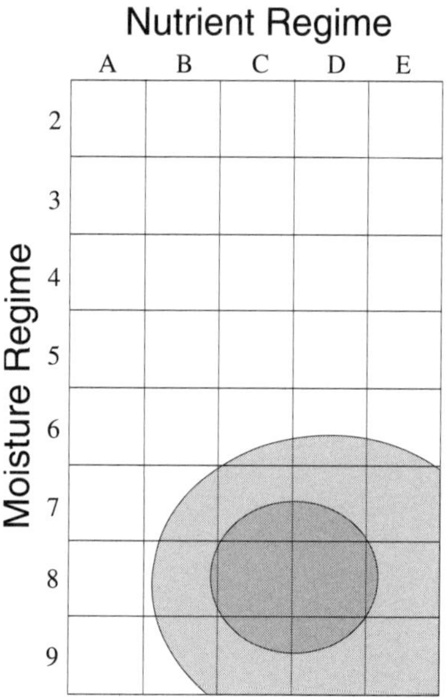

Tamarack is associated with the following ecosites:

k poor fen

l rich fen

aspen

tA

Populus tremuloides Michx.

Aspen (*Populus tremuloides*) is a deciduous, hardwood tree species. It grows under a variety of site conditions but grows best on well-drained, moist soils (Gimbarzevsky 1973; Johnson et al. 1995). Aspen grows in pure stands but is also commonly mixed with white spruce, balsam poplar, jack pine, or white birch (Farrar 1995). As aspen is not very shade-tolerant, it is usually dominant when mixed with other species (Sayn-Wittgenstein 1978).

Aspen has a full, rounded crown that, from above, appears small and round to irregular (Sayn-Wittgenstein 1978). The aspen crown sometimes appears "fuzzy" or "fluffy" due to its trembling leaves and almost never appears coarse or ragged in texture (Lutz and Caporaso 1959). On aerial photographs, aspen stands appear light gray to moderately light gray in tone and fine-grained or smooth in texture. Often, the ground can be seen through the canopy. Aspen cast light shadows and distinct individual trees can be recognized on aerial photographs.

Aspen is difficult to distinguish from white birch and balsam poplar on aerial photographs. White birch tends to be denser, brighter and has a less distinct perimeter (Sayn-Wittgenstein 1978). Also, aspen tends to be more sensitive than white birch to poor drainage (Lutz and Caporaso 1959) and therefore is not typically found in wetter areas. Although not always distinguishable, balsam poplar crowns are more pointed than those of aspen (Stoeckeler 1952; Lutz and Caporaso 1959), and balsam poplar is found on wetter sites than aspen.

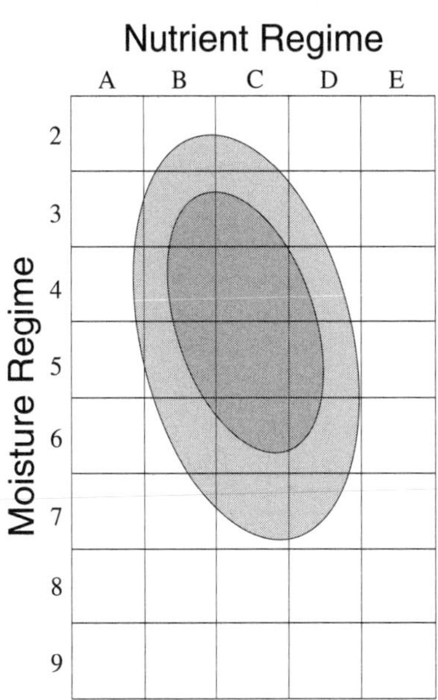

Aspen is associated with the following ecosites:

b blueberry

d low-bush cranberry

e dogwood

f ostrich fern

h horsetail

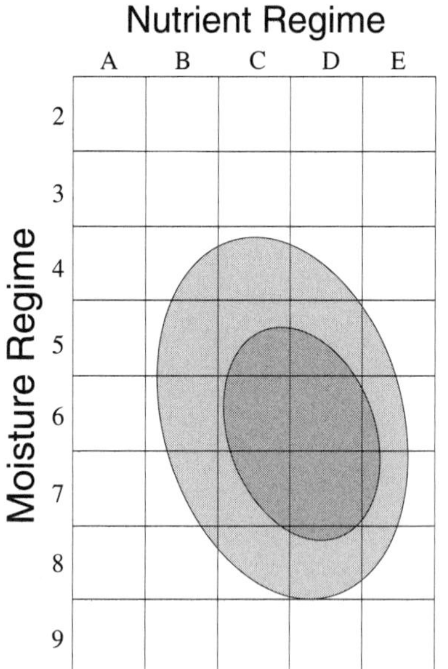

balsam poplar

bP

Populus balsamifera L.

Balsam poplar (*Populus balsamifera*) is a deciduous, hardwood tree species that is found on moist, rich, low-lying sites. Floodplains, terraces, stream banks, river banks, lake shores, and sand and gravel flats are some of the sites on which balsam poplar is found (Lutz and Caporaso 1959; Gimbarzevsky 1973; Sayn-Wittgenstein 1978; Farrar 1995; Johnson 1995). Alder (*Alnus* spp.), willow (*Salix* spp.), black spruce, balsam fir, and white spruce are often found in association with balsam poplar (Farrar 1995).

Balsam poplar is very similar in shape and appearance to aspen. Balsam poplar has a full, rounded crown that tends toward being conical and pointed. On aerial photographs, balsam poplar stands appear light gray to moderately light gray in tone and fine-grained or smooth in texture.

Balsam poplar is difficult to distinguish from aspen on aerial photographs. The only distinguishing features that can be used in identification are balsam poplar has a slightly more conical crown, which is difficult to see, and it is often found on wetter sites than aspen.

Balsam poplar is associated with the following ecosites:

d low-bush cranberry

e dogwood

f ostrich fern

h horsetail

white birch

wB

Betula papyrifera Marsh.

White birch (*Betula papyrifera*) is a deciduous, hardwood tree species that is shade intolerant. White birch has a wide distribution and is found primarily on well-drained but moist sites along forest edges, lake shores, and roadsides (Gimbarzevsky 1973; Farrar 1995; Johnson et al. 1995).

White birch usually grows in pure stands and mixes with aspen, balsam poplar, maple, jack pine, balsam fir and spruce, usually in clusters (Gimbarzevsky 1973; Sayn-Wittgenstein 1978; Farrar 1995). White birch typically grows on talus slopes or morainal deposits and is among the first species to reforest an area that has been burned or cut (Farrar 1995).

White birch is very similar to aspen and balsam poplar in shape and appearance; however, white birch tends to register as the lightest of the three on aerial photographs (Lutz and Caporaso 1959). Mature crowns are broad, rounded, irregular and compact, often divided into several main branches that are often crooked (Lutz and Caporaso 1959; Gimbarzevsky 1973; Sayn-Wittgenstein 1978). Stand texture is fine, silky, and smooth with small, conical crowns in younger stands but much coarser with rounded, irregular crowns of older stands (Lutz and Caporaso 1959; Sayn-Wittgenstein 1978). White birch trunks appear very white if observed from the side — even whiter than aspen, which are not perfectly white.

White birch is associated with the following ecosites:

b blueberry

d low-bush cranberry

e dogwood

f ostrich fern

h horsetail

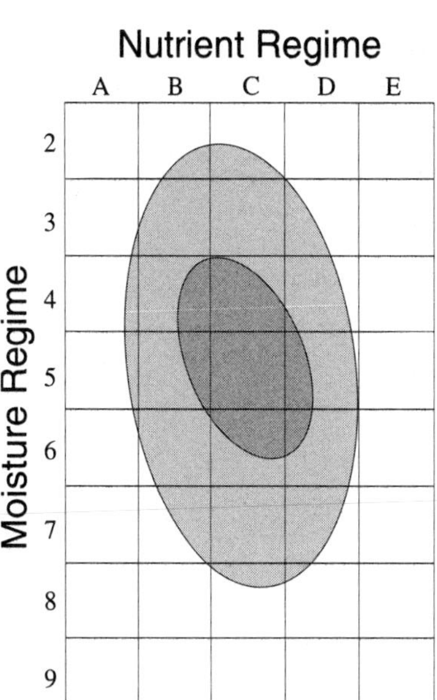

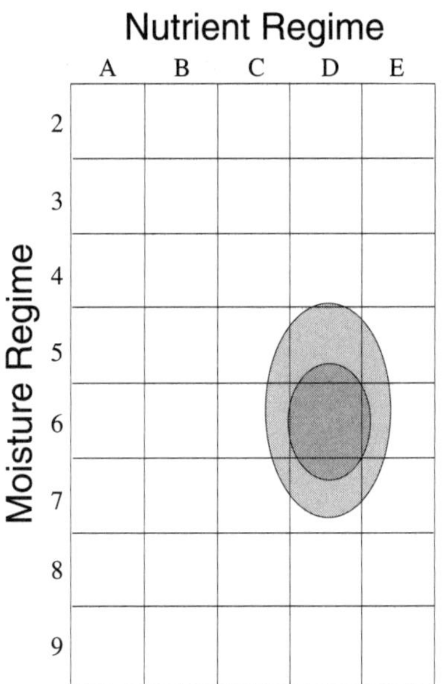

white elm

wE

Ulmus americana L.

White elm (*Ulmus americana*) is a deciduous, hardwood tree species that occurs mainly on alluvial flats where the water table is close to the surface and drainage is good (Farrar 1995; Johnson et al. 1995). White elm was exclusively found in association with balsam poplar, Manitoba maple, green ash, and white spruce, and it was sampled on the floodplains of the Mid-Boreal Lowland (MBL) ecoregion by Beckingham et al. (1996).

White elm, when growing in the open, is easily recognized by its large, broad, umbrella-shaped crown created by its large outwardly fanning branches divided close to the ground. In forested conditions, white elm develops a straight trunk, undivided for a considerable height (Sayn-Wittgenstein 1978).

White elm seldom forms pure stands, therefore, branching patterns are hard to see. Thus, white elm is difficult to separate from other hardwood species but it can be distinguished from aspen or white birch by its larger size (Sayn-Wittgenstein 1978).

White elm is associated with the following ecosite:

f ostrich fern

green ash

gA

Fraxinus pennsylvanica Marsh.

Green ash (*Fraxinus pennsylvanica*) is a deciduous, hardwood tree species preferring the moist, rich soils of bottomlands, lake shores, river valleys, and river banks (Farrar 1995; Johnson et al. 1995). Green ash is very successful in shelterbelts because it is hardy and fast-growing (Johnson et al. 1995).

Green ash, like all ash occurring in eastern Canada, rarely forms pure stands, either growing singularly or in a small group, has a straight trunk with little taper, and a medium sized, shallow, open crown (Sayn-Wittgenstein 1978).

Green ash is difficult to distinguish from other hardwoods by crown shape; however, the tone of ash foliage on aerial photographs is lighter than that of elm (Sayn-Wittgenstein 1978).

Green ash is associated with the following ecosite:

f ostrich fern

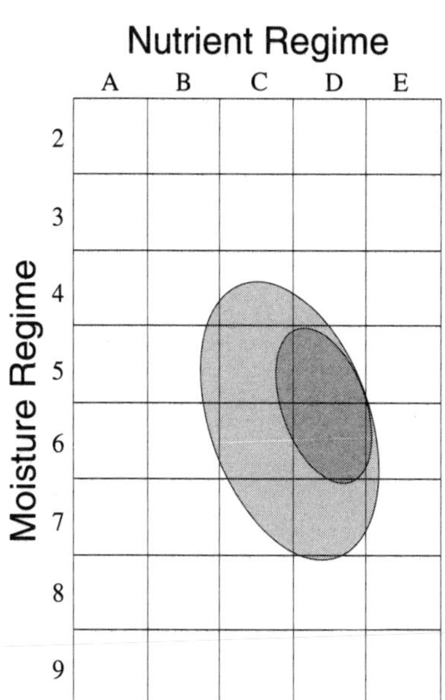

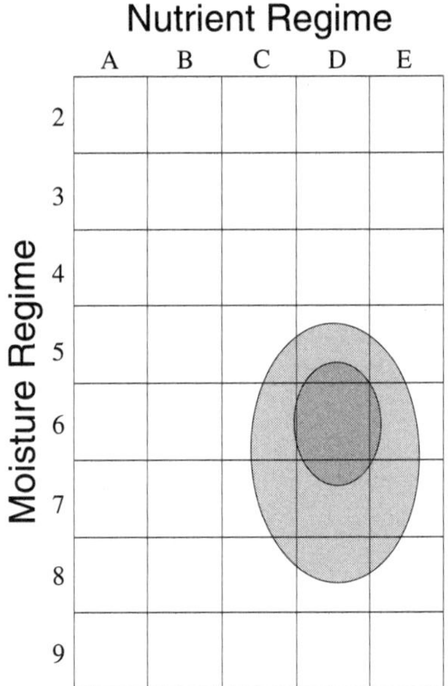

Manitoba maple

m M

Acer negundo L.

Manitoba maple (*Acer negundo*) is a deciduous, hardwood tree species intolerant of shade. It colonizes disturbed sites along lake shores, stream and river banks, ravines, wooded valleys, and areas that are seasonally flooded (Farrar 1995; Johnson et al. 1995). This species, along with white elm and green ash, was found only in the Mid-Boreal Lowland (MBL) ecoregion by Beckingham et al. (1996).

The branches of Manitoba maple are like those of the elm in that they are long and slender, dividing closer to the base in open areas and becoming branched higher up in forested areas. The foliage of Manitoba maple is very dense and because of its large smooth leaves it often displays tones lighter than elm (Sayn-Wittgenstein 1960).

Manitoba maple is associated with the following ecosite:

f ostrich fern

APPENDIX 5

TEXTURE CLASSES

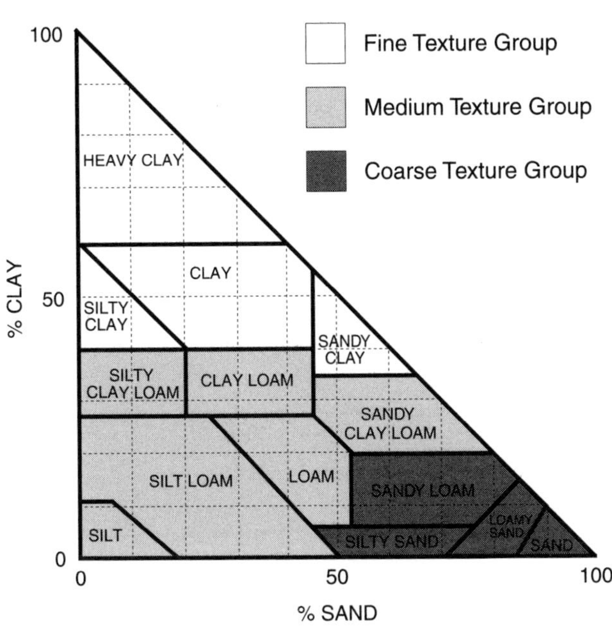

Texture Class	Forest Inventory[a]	Soil Survey[b]	Field Guide[c,e]
C	S, LS	S, LS	S, cS, mS, fS, LS, LcS, LmS, LfS
MC	SL, SiL	SL, fSL, (SL(g))[d]	SL, SiS
M	[f]	L, (L(g))[d], SiL, vfSL	[f]
MF	SCL, CL, SiCL	SCL, fSCL, vfSCL CL, SiCL	Si, SiL, L
F	SC, C, hC, SiC	SC, C, hC, SiC	SCL, CL, SiCL, SC, C, SiC, hC

[a] Lindenas, D.G. 1985. Specifications for the interpretation and mapping of aerial photographs in the forest inventory section. Saskatchewan Parks and Renew. Resourc. For. Div., Prince Albert, Saskatchewan. Intern. Doc.

[b] Ayres, K.W.; Anderson, D.W.; Ellis, J.G. 1978. The soils of the northern provincial forest in the Pasquia Hills and Saskatchewan portion of The Pas Map Area (63-E and 63-F Sask.). Sask. Inst. Pedol., Saskatoon, Saskatchewan. Publ. SF4.

[c] Beckingham, J.D.; Nielsen, D.G.; Futoransky, V.A. 1996. Field guide to ecosites of the mid-boreal ecoregions of Saskatchewan. Nat. Resour. Can., Can. For. Serv., Northwest Reg., North. For. Cent., Edmonton, Alberta. Spec. Rep. 6.

[d] Those textures presented in parenthesis occur on older soil survey maps.

[e] The Field Guide classes refer to Effective Texture Classes as follows: C = 1, MC = 2, MF = 3, and F = 4.

[f] The Medium Texture Class does not exist in forest inventory or field guide systems.

APPENDIX 7

SOIL ASSOCIATIONS

Dominantly Brunisolic and Regosolic soils	
Kk	Kewanoke
Pn	Pine
Sk	Sipanok
Dominantly Luvisolic soils	
Bt	Bittern LaKE
Bd	Bodmin
bo	Bow River
Do	Dorintosh
Lc	La Corne
Ln	Loon River
Pr	Piprell
Wv	Waitville
Wt	Waterhen
Dominantly Gleysolic soils	
Aw	Arbow
Mh	Marsh
Mw	Meadow
Miscellaneous mineral soils	
Av	Alluvium
Bx	Beach
Organic soils	
Bb	Bowl Bog
Bf	Flat Bog
Bl	Bagwa Lake
Bs	Stream Bog
Fb	Bowl Fen
Fh	Horizontal Fen
Fs	Stream Fen
Ll	Lavallee Lake

"BEST-FIT" ASSIGNMENT OF FOREST INVENTORY MAP UNITS[a] TO ECOSITE[b]

	INVENTORY MAINTENANCE MAP DATA			
MAIN CANOPY TREE SPECIES	MAP DRAINAGE	MAP TEXTURE	ECOSITE GENERAL	ECOSITE SECONDARY
Pure jack pine				
JP	VR, R, RW	C, CMC, MC	a	a
JP	W, WMW, MW, MWI	C, CMC, MC, MCMF, MF, MFF, F	c	c
JP	I, IP, P, PVP	C, CMC, MC, MCMF, MF, MFF, F, O	g	g
Pure black spruce				
BS	R, RW, W, WMW, MW	C, CMC, MC, MCMF, MF, MFF, F	c	c
BS	MWI, I, IP, P	C, CMC, MC, MCMF, MF, MFF, F	g	g
BS	P, PVP	O	j	j
Black spruce as a primary species without tL				
BSBF	R, RW, W, WMW, MW	C, CMC, MC	c	c
BSBFTA	R, RW, W, WMW, MW	C, CMC, MC	c	c
BSBFWB	R, RW, W, WMW, MW	C, CMC, MC	c	c
BSTA	R, RW, W, WMW, MW	C, CMC, MC	c	c
BSWB	R, RW, W, WMW, MW	C, CMC, MC	c	c
BSWS	R, RW, W, WMW, MW	C, CMC, MC	c	c
BSWSTA	R, RW, W, WMW, MW	C, CMC, MC	c	c
BSWSWB	R, RW, W, WMW, MW	C, CMC, MC	c	c
BSBF	W, WMW, MW	MCMF, MF, MFF, F	d	d
BSBFTA	W, WMW, MW	MCMF, MF, MFF, F	d	d
BSBFWB	W, WMW, MW	MCMF, MF, MFF, F	d	d
BSTA	W, WMW, MW	MCMF, MF, MFF, F	d	d
BSWB	W, WMW, MW	MCMF, MF, MFF, F	d	d
BSWS	W, WMW, MW	MCMF, MF, MFF, F	d	d
BSWSTA	W, WMW, MW	MCMF, MF, MFF, F	d	d
BSWSWB	W, WMW, MW	MCMF, MF, MFF, F	d	d
BSBF	MWI	C, CMC, MC	c–g	c–g
BSBFTA	MWI	C, CMC, MC	c–g	c–g
BSBFWB	MWI	C, CMC, MC	c–g	c–g
BSTA	MWI	C, CMC, MC	c–g	c–g
BSWB	MWI	C, CMC, MC	c–g	c–g
BSWS	MWI	C, CMC, MC	c–g	c–g
BSWSTA	MWI	C, CMC, MC	c–g	c–g
BSWSWB	MWI	C, CMC, MC	c–g	c–g
BSBF	MWI	MCMF, MF, MFF, F	d–h	d–h
BSBFTA	MWI	MCMF, MF, MFF, F	d–h	d–h
BSBFWB	MWI	MCMF, MF, MFF, F	d–h	d–h
BSTA	MWI	MCMF, MF, MFF, F	d–h	d–h
BSWB	MWI	MCMF, MF, MFF, F	d–h	d–h
BSWS	MWI	MCMF, MF, MFF, F	d–h	d–h
BSWSTA	MWI	MCMF, MF, MFF, F	d–h	d–h
BSWSWB	MWI	MCMF, MF, MFF, F	d–h	d–h
BSBF	I, IP, P	C, CMC, MC	g	g
BSBFTA	I, IP, P	C, CMC, MC	g	g
BSBFWB	I, IP, P	C, CMC, MC	g	g
BSTA	I, IP, P	C, CMC, MC	g	g
BSWB	I, IP, P	C, CMC, MC	g	g
BSWS	I, IP, P	C, CMC, MC	g	g
BSWSTA	I, IP, P	C, CMC, MC	g	g
BSWSWB	I, IP, P	C, CMC, MC	g	g

	INVENTORY MAINTENANCE MAP DATA			
MAIN CANOPY TREE SPECIES	MAP DRAINAGE	MAP TEXTURE	ECOSITE GENERAL	ECOSITE SECONDARY
BSBF	I, IP, P	MCMF, MF, MFF, F	h	h
BSBFTA	I, IP, P	MCMF, MF, MFF, F	h	h
BSBFWB	I, IP, P	MCMF, MF, MFF, F	h	h
BSTA	I, IP, P	MCMF, MF, MFF, F	h	h
BSWB	I, IP, P	MCMF, MF, MFF, F	h	h
BSWS	I, IP, P	MCMF, MF, MFF, F	h	h
BSWSTA	I, IP, P	MCMF, MF, MFF, F	h	h
BSWSWB	I, IP, P	MCMF, MF, MFF, F	h	h
BSBF	P, PVP	O	k	k
BSBFTA	P, PVP	O	k	k
BSBFWB	P, PVP	O	k	k
BSTA	P, PVP	O	k	k
BSWB	P, PVP	O	k	k
BSWS	P, PVP	O	k	k
BSWSTA	P, PVP	O	k	k
BSWSWB	P, PVP	O	k	k
Black spruce with tL and tA, wB or bP				
BSTLBP	W, WMW, MW	C, CMC, MC, MCMF, MF, MFF, F	d	d
BSTLTA	W, WMW, MW	C, CMC, MC, MCMF, MF, MFF, F	d	d
BSTLWB	W, WMW, MW	C, CMC, MC, MCMF, MF, MFF, F	d	d
BSTLBP	MWI, I, IP	C, CMC, MC, MCMF, MF, MFF, F	h	h
BSTLTA	MWI, I, IP	C, CMC, MC, MCMF, MF, MFF, F	h	h
BSTLWB	MWI, I, IP	C, CMC, MC, MCMF, MF, MFF, F	h	h
BSTLBP	P, PVP	C, CMC, MC, MCMF, MF, MFF, F, O	k	k
BSTLTA	P, PVP	C, CMC, MC, MCMF, MF, MFF, F, O	k	k
BSTLWB	P, PVP	C, CMC, MC, MCMF, MF, MFF, F, O	k	k
Black spruce with tL as a secondary species				
BSTL	W, WMW, MW	C, CMC, MC, MCMF, MF, MFF, F	d	d
BSTL	MWI, I	C, CMC, MC, MCMF, MF, MFF, F	h	h
BSTL	IP, P, PVP	C, CMC, MC, MCMF, MF, MFF, F, O	k	k
Jack pine with black spruce				
JPBS	R, RW, W, WMW, MW	C, CMC, MC, MCMF, MF, MFF, F	c	c
JPBS	MWI	C, CMC, MC, MCMF, MF, MFF, F	c–g	c–g
JPBS	I, IP, P	C, CMC, MC, MCMF, MF, MFF, F	g	g
JPBS	P, VP	O	j	j
Black spruce with jack pine				
BSJP	R, RW, W, WMW, MW	C, CMC, MC, MCMF, MF, MFF, F	c	c
BSJP	MWI	C, CMC, MC, MCMF, MF, MFF, F	c–g	c–g
BSJP	I, IP, P	C, CMC, MC, MCMF, MF, MFF, F	g	g
BSJP	P, VP	O	j	j
Jack pine with tamarack				
JPTL	PVP	O	k	k
Jack pine, aspen and birch mixedwoods				
JPTA	RW, W	C, CMC, MC, MCMF, MF, MFF, F	b	b
JPWB	RW, W	C, CMC, MC, MCMF, MF, MFF, F	b	b
TAJP	RW, W	C, CMC, MC, MCMF, MF, MFF, F	b	b
TAWBJP	RW, W	C, CMC, MC, MCMF, MF, MFF, F	b	b
WBJP	RW, W	C, CMC, MC, MCMF, MF, MFF, F	b	b
WBTAJP	RW, W	C, CMC, MC, MCMF, MF, MFF, F	b	b
JPTA	WMW	C, CMC, MC, MCMF, MF, MFF, F	b–d	b–d

	INVENTORY MAINTENANCE MAP DATA			
MAIN CANOPY TREE SPECIES	MAP DRAINAGE	MAP TEXTURE	ECOSITE GENERAL	ECOSITE SECONDARY
JPWB	WMW	C, CMC, MC, MCMF, MF, MFF, F	b–d	b–d
TAJP	WMW	C, CMC, MC, MCMF, MF, MFF, F	b–d	b–d
TAWBJP	WMW	C, CMC, MC, MCMF, MF, MFF, F	b–d	b–d
WBJP	WMW	C, CMC, MC, MCMF, MF, MFF, F	b–d	b–d
WBTAJP	WMW	C, CMC, MC, MCMF, MF, MFF, F	b–d	b–d
JPTA	MW	C, CMC, MC, MCMF, MF, MFF, F	d	d
JPWB	MW	C, CMC, MC, MCMF, MF, MFF, F	d	d
TAJP	MW	C, CMC, MC, MCMF, MF, MFF, F	d	d
TAWBJP	MW	C, CMC, MC, MCMF, MF, MFF, F	d	d
WBJP	MW	C, CMC, MC, MCMF, MF, MFF, F	d	d
WBTAJP	MW	C, CMC, MC, MCMF, MF, MFF, F	d	d
JPTA	MWI	C, CMC, MC, MCMF, MF, MFF, F	d–h	d–h
JPWB	MWI	C, CMC, MC, MCMF, MF, MFF, F	d–h	d–h
TAJP	MWI	C, CMC, MC, MCMF, MF, MFF, F	d–h	d–h
TAWBJP	MWI	C, CMC, MC, MCMF, MF, MFF, F	d–h	d–h
WBJP	MWI	C, CMC, MC, MCMF, MF, MFF, F	d–h	d–h
WBTAJP	MWI	C, CMC, MC, MCMF, MF, MFF, F	d–h	d–h
JPTA	I, IP, P, PVP	C, CMC, MC, MCMF, MF, MFF, F, O	h	h
JPWB	I, IP, P, PVP	C, CMC, MC, MCMF, MF, MFF, F, O	h	h
TAJP	I, IP, P, PVP	C, CMC, MC, MCMF, MF, MFF, F, O	h	h
TAWBJP	I, IP, P, PVP	C, CMC, MC, MCMF, MF, MFF, F, O	h	h
WBJP	I, IP, P, PVP	C, CMC, MC, MCMF, MF, MFF, F, O	h	h
WBTAJP	I, IP, P, PVP	C, CMC, MC, MCMF, MF, MFF, F, O	h	h
Aspen, birch, balsam fir, white spruce mixtures				
BF	RW, W	C, CMC, MC	b	b
BFTA	RW, W	C, CMC, MC	b	b
BFWB	RW, W	C, CMC, MC	b	b
BFWS	RW, W	C, CMC, MC	b	b
BFWSTA	RW, W	C, CMC, MC	b	b
BFWSWB	RW, W	C, CMC, MC	b	b
TA	RW, W	C, CMC, MC	b	b
TABF	RW, W	C, CMC, MC	b	b
TAWB	RW, W	C, CMC, MC	b	b
TAWBBF	RW, W	C, CMC, MC	b	b
TAWBWS	RW, W	C, CMC, MC	b	b
TAWS	RW, W	C, CMC, MC	b	b
WB	RW, W	C, CMC, MC	b	b
WBBF	RW, W	C, CMC, MC	b	b
WBTA	RW, W	C, CMC, MC	b	b
WBTAWS	RW, W	C, CMC, MC	b	b
WBWS	RW, W	C, CMC, MC	b	b
WS	RW, W	C, CMC, MC	b	b
WSBF	RW, W	C, CMC, MC	b	b
WSBFTA	RW, W	C, CMC, MC	b	b
WSBFWB	RW, W	C, CMC, MC	b	b
WSTA	RW, W	C, CMC, MC	b	b
WSWB	RW, W	C, CMC, MC	b	b
BF	W	MCMF	b–d	b–d
BFTA	W	MCMF	b–d	b–d
BFWB	W	MCMF	b–d	b–d
BFWS	W	MCMF	b–d	b–d
BFWSTA	W	MCMF	b–d	b–d
BFWSWB	W	MCMF	b–d	b–d
TA	W	MCMF	b–d	b–d
TABF	W	MCMF	b–d	b–d

INVENTORY MAINTENANCE MAP DATA				
MAIN CANOPY TREE SPECIES	MAP DRAINAGE	MAP TEXTURE	ECOSITE GENERAL	ECOSITE SECONDARY
TAWB	W	MCMF	b–d	b–d
TAWBBF	W	MCMF	b–d	b–d
TAWBWS	W	MCMF	b–d	b–d
TAWS	W	MCMF	b–d	b–d
WB	W	MCMF	b–d	b–d
WBBF	W	MCMF	b–d	b–d
WBTA	W	MCMF	b–d	b–d
WBTAWS	W	MCMF	b–d	b–d
WBWS	W	MCMF	b–d	b–d
WS	W	MCMF	b–d	b–d
WSBF	W	MCMF	b–d	b–d
WSBFTA	W	MCMF	b–d	b–d
WSBFWB	W	MCMF	b–d	b–d
WSTA	W	MCMF	b–d	b–d
WSWB	W	MCMF	b–d	b–d
BF	W	MF, MFF, F	d	d
BFTA	W	MF, MFF, F	d	d
BFWB	W	MF, MFF, F	d	d
BFWS	W	MF, MFF, F	d	d
BFWSTA	W	MF, MFF, F	d	d
BFWSWB	W	MF, MFF, F	d	d
TA	W	MF, MFF, F	d	d
TABF	W	MF, MFF, F	d	d
TAWB	W	MF, MFF, F	d	d
TAWBBF	W	MF, MFF, F	d	d
TAWBWS	W	MF, MFF, F	d	d
TAWS	W	MF, MFF, F	d	d
WB	W	MF, MFF, F	d	d
WBBF	W	MF, MFF, F	d	d
WBTA	W	MF, MFF, F	d	d
WBTAWS	W	MF, MFF, F	d	d
WBWS	W	MF, MFF, F	d	d
WS	W	MF, MFF, F	d	d
WSBF	W	MF, MFF, F	d	d
WSBFTA	W	MF, MFF, F	d	d
WSBFWB	W	MF, MFF, F	d	d
WSTA	W	MF, MFF, F	d	d
WSWB	W	MF, MFF, F	d	d
BF	WMW	C, CMC, MC	b–d	b–d
BFTA	WMW	C, CMC, MC	b–d	b–d
BFWB	WMW	C, CMC, MC	b–d	b–d
BFWS	WMW	C, CMC, MC	b–d	b–d
BFWSTA	WMW	C, CMC, MC	b–d	b–d
BFWSWB	WMW	C, CMC, MC	b–d	b–d
TA	WMW	C, CMC, MC	b–d	b–d
TABF	WMW	C, CMC, MC	b–d	b–d
TAWB	WMW	C, CMC, MC	b–d	b–d
TAWBBF	WMW	C, CMC, MC	b–d	b–d
TAWBWS	WMW	C, CMC, MC	b–d	b–d
TAWS	WMW	C, CMC, MC	b–d	b–d
WB	WMW	C, CMC, MC	b–d	b–d
WBBF	WMW	C, CMC, MC	b–d	b–d
WBTA	WMW	C, CMC, MC	b–d	b–d
WBTAWS	WMW	C, CMC, MC	b–d	b–d
WBWS	WMW	C, CMC, MC	b–d	b–d
WS	WMW	C, CMC, MC	b–d	b–d

	INVENTORY MAINTENANCE MAP DATA			
MAIN CANOPY TREE SPECIES	MAP DRAINAGE	MAP TEXTURE	ECOSITE GENERAL	ECOSITE SECONDARY
WSBF	WMW	C, CMC, MC	b–d	b–d
WSBFTA	WMW	C, CMC, MC	b–d	b–d
WSBFWB	WMW	C, CMC, MC	b–d	b–d
WSTA	WMW	C, CMC, MC	b–d	b–d
WSWB	WMW	C, CMC, MC	b–d	b–d
BF	WMW	MCMF	d	d
BFTA	WMW	MCMF	d	d
BFWB	WMW	MCMF	d	d
BFWS	WMW	MCMF	d	d
BFWSTA	WMW	MCMF	d	d
BFWSWB	WMW	MCMF	d	d
TA	WMW	MCMF	d	d
TABF	WMW	MCMF	d	d
TAWB	WMW	MCMF	d	d
TAWBBF	WMW	MCMF	d	d
TAWBWS	WMW	MCMF	d	d
TAWS	WMW	MCMF	d	d
WB	WMW	MCMF	d	d
WBBF	WMW	MCMF	d	d
WBTA	WMW	MCMF	d	d
WBTAWS	WMW	MCMF	d	d
WBWS	WMW	MCMF	d	d
WS	WMW	MCMF	d	d
WSBF	WMW	MCMF	d	d
WSBFTA	WMW	MCMF	d	d
WSBFWB	WMW	MCMF	d	d
WSTA	WMW	MCMF	d	d
WSWB	WMW	MCMF	d	d
BF	WMW	MF, MFF, F	d	d
BFTA	WMW	MF, MFF, F	d	d
BFWB	WMW	MF, MFF, F	d	d
BFWS	WMW	MF, MFF, F	d	d
BFWSTA	WMW	MF, MFF, F	d	d
BFWSWB	WMW	MF, MFF, F	d	d
TA	WMW	MF, MFF, F	d	d
TABF	WMW	MF, MFF, F	d	d
TAWB	WMW	MF, MFF, F	d	d
TAWBBF	WMW	MF, MFF, F	d	d
TAWBWS	WMW	MF, MFF, F	d	d
TAWS	WMW	MF, MFF, F	d	d
WB	WMW	MF, MFF, F	d	d
WBBF	WMW	MF, MFF, F	d	d
WBTA	WMW	MF, MFF, F	d	d
WBTAWS	WMW	MF, MFF, F	d	d
WBWS	WMW	MF, MFF, F	d	d
WS	WMW	MF, MFF, F	d	d
WSBF	WMW	MF, MFF, F	d	d
WSBFTA	WMW	MF, MFF, F	d	d
WSBFWB	WMW	MF, MFF, F	d	d
WSTA	WMW	MF, MFF, F	d	d
WSWB	WMW	MF, MFF, F	d	d
BF	MW	C, CMC, MC, MCMF	d	d
BFTA	MW	C, CMC, MC, MCMF	d	d
BFWB	MW	C, CMC, MC, MCMF	d	d
BFWS	MW	C, CMC, MC, MCMF	d	d
BFWSTA	MW	C, CMC, MC, MCMF	d	d

	INVENTORY MAINTENANCE MAP DATA			
MAIN CANOPY TREE SPECIES	MAP DRAINAGE	MAP TEXTURE	ECOSITE GENERAL	ECOSITE SECONDARY
BFWSWB	MW	C, CMC, MC, MCMF	d	d
TA	MW	C, CMC, MC, MCMF	d	d
TABF	MW	C, CMC, MC, MCMF	d	d
TAWB	MW	C, CMC, MC, MCMF	d	d
TAWBBF	MW	C, CMC, MC, MCMF	d	d
TAWBWS	MW	C, CMC, MC, MCMF	d	d
TAWS	MW	C, CMC, MC, MCMF	d	d
WB	MW	C, CMC, MC, MCMF	d	d
WBBF	MW	C, CMC, MC, MCMF	d	d
WBTA	MW	C, CMC, MC, MCMF	d	d
WBTAWS	MW	C, CMC, MC, MCMF	d	d
WBWS	MW	C, CMC, MC, MCMF	d	d
WS	MW	C, CMC, MC, MCMF	d	d
WSBF	MW	C, CMC, MC, MCMF	d	d
WSBFTA	MW	C, CMC, MC, MCMF	d	d
WSBFWB	MW	C, CMC, MC, MCMF	d	d
WSTA	MW	C, CMC, MC, MCMF	d	d
WSWB	MW	C, CMC, MC, MCMF	d	d
BF	MW	MF, MFF, F	d	d
BFTA	MW	MF, MFF, F	d	d
BFWB	MW	MF, MFF, F	d	d
BFWS	MW	MF, MFF, F	d	d
BFWSTA	MW	MF, MFF, F	d	d
BFWSWB	MW	MF, MFF, F	d	d
TA	MW	MF, MFF, F	d	d
TABF	MW	MF, MFF, F	d	d
TAWB	MW	MF, MFF, F	d	d
TAWBBF	MW	MF, MFF, F	d	d
TAWBWS	MW	MF, MFF, F	d	d
TAWS	MW	MF, MFF, F	d	d
WB	MW	MF, MFF, F	d	d
WBBF	MW	MF, MFF, F	d	d
WBTA	MW	MF, MFF, F	d	d
WBTAWS	MW	MF, MFF, F	d	d
WBWS	MW	MF, MFF, F	d	d
WS	MW	MF, MFF, F	d	d
WSBF	MW	MF, MFF, F	d	d
WSBFTA	MW	MF, MFF, F	d	d
WSBFWB	MW	MF, MFF, F	d	d
WSTA	MW	MF, MFF, F	d	d
WSWB	MW	MF, MFF, F	d	d
BF	MWI	C, CMC, MC, MCMF	d–e	d–e
BFTA	MWI	C, CMC, MC, MCMF	d–e	d–e
BFWB	MWI	C, CMC, MC, MCMF	d–e	d–e
BFWS	MWI	C, CMC, MC, MCMF	d–e	d–e
BFWSTA	MWI	C, CMC, MC, MCMF	d–e	d–e
BFWSWB	MWI	C, CMC, MC, MCMF	d–e	d–e
TA	MWI	C, CMC, MC, MCMF	d–e	d–e
TABF	MWI	C, CMC, MC, MCMF	d–e	d–e
TAWB	MWI	C, CMC, MC, MCMF	d–e	d–e
TAWBBF	MWI	C, CMC, MC, MCMF	d–e	d–e
TAWBWS	MWI	C, CMC, MC, MCMF	d–e	d–e
TAWS	MWI	C, CMC, MC, MCMF	d–e	d–e
WB	MWI	C, CMC, MC, MCMF	d–e	d–e
WBBF	MWI	C, CMC, MC, MCMF	d–e	d–e
WBTA	MWI	C, CMC, MC, MCMF	d–e	d–e

	INVENTORY MAINTENANCE MAP DATA			
MAIN CANOPY TREE SPECIES	MAP DRAINAGE	MAP TEXTURE	ECOSITE GENERAL	ECOSITE SECONDARY
WBTAWS	MWI	C, CMC, MC, MCMF	d–e	d–e
WBWS	MWI	C, CMC, MC, MCMF	d–e	d–e
WS	MWI	C, CMC, MC, MCMF	d–e	d–e
WSBF	MWI	C, CMC, MC, MCMF	d–e	d–e
WSBFTA	MWI	C, CMC, MC, MCMF	d–e	d–e
WSBFWB	MWI	C, CMC, MC, MCMF	d–e	d–e
WSTA	MWI	C, CMC, MC, MCMF	d–e	d–e
WSWB	MWI	C, CMC, MC, MCMF	d–e	d–e
BF	MWI	MF, MFF, F	d–e	d–e
BFTA	MWI	MF, MFF, F	d–e	d–e
BFWB	MWI	MF, MFF, F	d–e	d–e
BFWS	MWI	MF, MFF, F	d–e	d–e
BFWSTA	MWI	MF, MFF, F	d–e	d–e
BFWSWB	MWI	MF, MFF, F	d–e	d–e
TA	MWI	MF, MFF, F	d–e	d–e
TABF	MWI	MF, MFF, F	d–e	d–e
TAWB	MWI	MF, MFF, F	d–e	d–e
TAWBBF	MWI	MF, MFF, F	d–e	d–e
TAWBWS	MWI	MF, MFF, F	d–e	d–e
TAWS	MWI	MF, MFF, F	d–e	d–e
WB	MWI	MF, MFF, F	d–e	d–e
WBBF	MWI	MF, MFF, F	d–e	d–e
WBTA	MWI	MF, MFF, F	d–e	d–e
WBTAWS	MWI	MF, MFF, F	d–e	d–e
WBWS	MWI	MF, MFF, F	d–e	d–e
WS	MWI	MF, MFF, F	d–e	d–e
WSBF	MWI	MF, MFF, F	d–e	d–e
WSBFTA	MWI	MF, MFF, F	d–e	d–e
WSBFWB	MWI	MF, MFF, F	d–e	d–e
WSTA	MWI	MF, MFF, F	d–e	d–e
WSWB	MWI	MF, MFF, F	d–e	d–e
BF	I	C, CMC, MC, MCMF, MF, MFF, F	e	e
BFTA	I	C, CMC, MC, MCMF, MF, MFF, F	e	e
BFWB	I	C, CMC, MC, MCMF, MF, MFF, F	e	e
BFWS	I	C, CMC, MC, MCMF, MF, MFF, F	e	e
BFWSTA	I	C, CMC, MC, MCMF, MF, MFF, F	e	e
BFWSWB	I	C, CMC, MC, MCMF, MF, MFF, F	e	e
TA	I	C, CMC, MC, MCMF, MF, MFF, F	e	e
TABF	I	C, CMC, MC, MCMF, MF, MFF, F	e	e
TAWB	I	C, CMC, MC, MCMF, MF, MFF, F	e	e
TAWBBF	I	C, CMC, MC, MCMF, MF, MFF, F	e	e
TAWBWS	I	C, CMC, MC, MCMF, MF, MFF, F	e	e
TAWS	I	C, CMC, MC, MCMF, MF, MFF, F	e	e
WB	I	C, CMC, MC, MCMF, MF, MFF, F	e	e
WBBF	I	C, CMC, MC, MCMF, MF, MFF, F	e	e
WBTA	I	C, CMC, MC, MCMF, MF, MFF, F	e	e
WBTAWS	I	C, CMC, MC, MCMF, MF, MFF, F	e	e
WBWS	I	C, CMC, MC, MCMF, MF, MFF, F	e	e
WS	I	C, CMC, MC, MCMF, MF, MFF, F	e	e
WSBF	I	C, CMC, MC, MCMF, MF, MFF, F	e	e
WSBFTA	I	C, CMC, MC, MCMF, MF, MFF, F	e	e
WSBFWB	I	C, CMC, MC, MCMF, MF, MFF, F	e	e
WSTA	I	C, CMC, MC, MCMF, MF, MFF, F	e	e
WSWB	I	C, CMC, MC, MCMF, MF, MFF, F	e	e
BF	IP	C, CMC, MC, MCMF, MF, MFF, F	e–h	e–h
BFTA	IP	C, CMC, MC, MCMF, MF, MFF, F	e–h	e–h

	INVENTORY MAINTENANCE MAP DATA			
MAIN CANOPY TREE SPECIES	MAP DRAINAGE	MAP TEXTURE	ECOSITE GENERAL	ECOSITE SECONDARY
BFWB	IP	C, CMC, MC, MCMF, MF, MFF, F	e–h	e–h
BFWS	IP	C, CMC, MC, MCMF, MF, MFF, F	e–h	e–h
BFWSTA	IP	C, CMC, MC, MCMF, MF, MFF, F	e–h	e–h
BFWSWB	IP	C, CMC, MC, MCMF, MF, MFF, F	e–h	e–h
TA	IP	C, CMC, MC, MCMF, MF, MFF, F	e–h	e–h
TABF	IP	C, CMC, MC, MCMF, MF, MFF, F	e–h	e–h
TAWB	IP	C, CMC, MC, MCMF, MF, MFF, F	e–h	e–h
TAWBBF	IP	C, CMC, MC, MCMF, MF, MFF, F	e–h	e–h
TAWBWS	IP	C, CMC, MC, MCMF, MF, MFF, F	e–h	e–h
TAWS	IP	C, CMC, MC, MCMF, MF, MFF, F	e–h	e–h
WB	IP	C, CMC, MC, MCMF, MF, MFF, F	e–h	e–h
WBBF	IP	C, CMC, MC, MCMF, MF, MFF, F	e–h	e–h
WBTA	IP	C, CMC, MC, MCMF, MF, MFF, F	e–h	e–h
WBTAWS	IP	C, CMC, MC, MCMF, MF, MFF, F	e–h	e–h
WBWS	IP	C, CMC, MC, MCMF, MF, MFF, F	e–h	e–h
WS	IP	C, CMC, MC, MCMF, MF, MFF, F	e–h	e–h
WSBF	IP	C, CMC, MC, MCMF, MF, MFF, F	e–h	e–h
WSBFTA	IP	C, CMC, MC, MCMF, MF, MFF, F	e–h	e–h
WSBFWB	IP	C, CMC, MC, MCMF, MF, MFF, F	e–h	e–h
WSTA	IP	C, CMC, MC, MCMF, MF, MFF, F	e–h	e–h
WSWB	IP	C, CMC, MC, MCMF, MF, MFF, F	e–h	e–h
BF	P, PVP	C, CMC, MC, MCMF, MF, MFF, F, O	h	h
BFTA	P, PVP	C, CMC, MC, MCMF, MF, MFF, F, O	h	h
BFWB	P, PVP	C, CMC, MC, MCMF, MF, MFF, F, O	h	h
BFWS	P, PVP	C, CMC, MC, MCMF, MF, MFF, F, O	h	h
BFWSTA	P, PVP	C, CMC, MC, MCMF, MF, MFF, F, O	h	h
BFWSWB	P, PVP	C, CMC, MC, MCMF, MF, MFF, F, O	h	h
TA	P, PVP	C, CMC, MC, MCMF, MF, MFF, F, O	h	h
TABF	P, PVP	C, CMC, MC, MCMF, MF, MFF, F, O	h	h
TAWB	P, PVP	C, CMC, MC, MCMF, MF, MFF, F, O	h	h
TAWBBF	P, PVP	C, CMC, MC, MCMF, MF, MFF, F, O	h	h
TAWBWS	P, PVP	C, CMC, MC, MCMF, MF, MFF, F, O	h	h
TAWS	P, PVP	C, CMC, MC, MCMF, MF, MFF, F, O	h	h
WB	P, PVP	C, CMC, MC, MCMF, MF, MFF, F, O	h	h
WBBF	P, PVP	C, CMC, MC, MCMF, MF, MFF, F, O	h	h
WBTA	P, PVP	C, CMC, MC, MCMF, MF, MFF, F, O	h	h
WBTAWS	P, PVP	C, CMC, MC, MCMF, MF, MFF, F, O	h	h
WBWS	P, PVP	C, CMC, MC, MCMF, MF, MFF, F, O	h	h
WS	P, PVP	C, CMC, MC, MCMF, MF, MFF, F, O	h	h
WSBF	P, PVP	C, CMC, MC, MCMF, MF, MFF, F, O	h	h
WSBFTA	P, PVP	C, CMC, MC, MCMF, MF, MFF, F, O	h	h
WSBFWB	P, PVP	C, CMC, MC, MCMF, MF, MFF, F, O	h	h
WSTA	P, PVP	C, CMC, MC, MCMF, MF, MFF, F, O	h	h
WSWB	P, PVP	C, CMC, MC, MCMF, MF, MFF, F, O	h	h
Black spruce, tA, wS, bP, jP and wB mixtures				
TABS	RW, W	C, CMC, MC	b	b
TAWBBS	RW, W	C, CMC, MC	b	b
WBBS	RW, W	C, CMC, MC	b	b
WBTABS	RW, W	C, CMC, MC	b	b
WSBS	RW, W	C, CMC, MC	b	b
WSBSTA	RW, W	C, CMC, MC	b	b
WSBSWB	RW, W	C, CMC, MC	b	b
WSJPBP	RW, W	C, CMC, MC	b	b
TABS	W	MCMF	b–d	b–d
TAWBBS	W	MCMF	b–d	b–d

	INVENTORY MAINTENANCE MAP DATA			
MAIN CANOPY TREE SPECIES	MAP DRAINAGE	MAP TEXTURE	ECOSITE GENERAL	ECOSITE SECONDARY
WBBS	W	MCMF	b–d	b–d
WBTABS	W	MCMF	b–d	b–d
WSBS	W	MCMF	b–d	b–d
WSBSTA	W	MCMF	b–d	b–d
WSBSWB	W	MCMF	b–d	b–d
WSJPBP	W	MCMF	b–d	b–d
TABS	W	MF, MFFMF	d	d
TAWBBS	W	MF, MFFMF	d	d
WBBS	W	MF, MFFMF	d	d
WBTABS	W	MF, MFFMF	d	d
WSBS	W	MF, MFFMF	d	d
WSBSTA	W	MF, MFFMF	d	d
WSBSWB	W	MF, MFFMF	d	d
WSJPBP	W	MF, MFFMF	d	d
TABS	WMW	C, CMC, MC	b–d	b–d
TAWBBS	WMW	C, CMC, MC	b–d	b–d
WBBS	WMW	C, CMC, MC	b–d	b–d
WBTABS	WMW	C, CMC, MC	b–d	b–d
WSBS	WMW	C, CMC, MC	b–d	b–d
WSBSTA	WMW	C, CMC, MC	b–d	b–d
WSBSWB	WMW	C, CMC, MC	b–d	b–d
WSJPBP	WMW	C, CMC, MC	b–d	b–d
TABS	WMW	MCMF, MF, MFF, F	d	d
TAWBBS	WMW	MCMF, MF, MFF, F	d	d
WBBS	WMW	MCMF, MF, MFF, F	d	d
WBTABS	WMW	MCMF, MF, MFF, F	d	d
WSBS	WMW	MCMF, MF, MFF, F	d	d
WSBSTA	WMW	MCMF, MF, MFF, F	d	d
WSBSWB	WMW	MCMF, MF, MFF, F	d	d
WSJPBP	WMW	MCMF, MF, MFF, F	d	d
TABS	MW	C, CMC, MC, MCMF, MF, MFF, F	d	d
TAWBBS	MW	C, CMC, MC, MCMF, MF, MFF, F	d	d
WBBS	MW	C, CMC, MC, MCMF, MF, MFF, F	d	d
WBTABS	MW	C, CMC, MC, MCMF, MF, MFF, F	d	d
WSBS	MW	C, CMC, MC, MCMF, MF, MFF, F	d	d
WSBSTA	MW	C, CMC, MC, MCMF, MF, MFF, F	d	d
WSBSWB	MW	C, CMC, MC, MCMF, MF, MFF, F	d	d
WSJPBP	MW	C, CMC, MC, MCMF, MF, MFF, F	d	d
TABS	MWI	C, CMC, MC, MCMF, MF, MFF, F	d–h	d–h
TAWBBS	MWI	C, CMC, MC, MCMF, MF, MFF, F	d–h	d–h
WBBS	MWI	C, CMC, MC, MCMF, MF, MFF, F	d–h	d–h
WBTABS	MWI	C, CMC, MC, MCMF, MF, MFF, F	d–h	d–h
WSBS	MWI	C, CMC, MC, MCMF, MF, MFF, F	d–h	d–h
WSBSTA	MWI	C, CMC, MC, MCMF, MF, MFF, F	d–h	d–h
WSBSWB	MWI	C, CMC, MC, MCMF, MF, MFF, F	d–h	d–h
WSJPBP	MWI	C, CMC, MC, MCMF, MF, MFF, F	d–h	d–h
TABS	I, IP, P, PVP	C, CMC, MC, MCMF, MF, MFF, F, O	h	h
TAWBBS	I, IP, P, PVP	C, CMC, MC, MCMF, MF, MFF, F, O	h	h
WBBS	I, IP, P, PVP	C, CMC, MC, MCMF, MF, MFF, F, O	h	h
WBTABS	I, IP, P, PVP	C, CMC, MC, MCMF, MF, MFF, F, O	h	h
WSBS	I, IP, P, PVP	C, CMC, MC, MCMF, MF, MFF, F, O	h	h
WSBSTA	I, IP, P, PVP	C, CMC, MC, MCMF, MF, MFF, F, O	h	h
WSBSWB	I, IP, P, PVP	C, CMC, MC, MCMF, MF, MFF, F, O	h	h
WSJPBP	I, IP, P, PVP	C, CMC, MC, MCMF, MF, MFF, F, O	h	h

	INVENTORY MAINTENANCE MAP DATA			
MAIN CANOPY TREE SPECIES	MAP DRAINAGE	MAP TEXTURE	ECOSITE GENERAL	ECOSITE SECONDARY
White spruce, jP, tA, and wB mixtures				
WSJP	RW, W	C, CMC, MC	b	b
WSJPTA	RW, W	C, CMC, MC	b	b
WSJPWB	RW, W	C, CMC, MC	b	b
WSJP	W	MCMF	b–d	b–d
WSJPTA	W	MCMF	b–d	b–d
WSJPWB	W	MCMF	b–d	b–d
WSJP	W	MF, MFF, F	d	d
WSJPTA	W	MF, MFF, F	d	d
WSJPWB	W	MF, MFF, F	d	d
WSJP	WMW	C, CMC, MC	b–d	b–d
WSJPTA	WMW	C, CMC, MC	b–d	b–d
WSJPWB	WMW	C, CMC, MC	b–d	b–d
WSJP	WMW	MCMF, MF, MFF, F	d	d
WSJPTA	WMW	MCMF, MF, MFF, F	d	d
WSJPWB	WMW	MCMF, MF, MFF, F	d	d
WSJP	MW	C, CMC, MC, MCMF, MF, MFF, F	d	d
WSJPTA	MW	C, CMC, MC, MCMF, MF, MFF, F	d	d
WSJPWB	MW	C, CMC, MC, MCMF, MF, MFF, F	d	d
WSJP	MWI	C, CMC, MC, MCMF, MF, MFF, F	d–h	d–h
WSJPTA	MWI	C, CMC, MC, MCMF, MF, MFF, F	d–h	d–h
WSJPWB	MWI	C, CMC, MC, MCMF, MF, MFF, F	d–h	d–h
WSJP	I, IP, P, PVP	C, CMC, MC, MCMF, MF, MFF, F, O	h	h
WSJPTA	I, IP, P, PVP	C, CMC, MC, MCMF, MF, MFF, F, O	h	h
WSJPWB	I, IP, P, PVP	C, CMC, MC, MCMF, MF, MFF, F, O	h	h
Jack pine, wS, tA, wB mixtures				
JPWS	RW, W	C, CMC, MC, MCMF, MF, MFF, F	b	b
JPWSBP	RW, W	C, CMC, MC, MCMF, MF, MFF, F	b	b
JPWSTA	RW, W	C, CMC, MC, MCMF, MF, MFF, F	b	b
JPWSWB	RW, W	C, CMC, MC, MCMF, MF, MFF, F	b	b
JPWS	WMW	C, CMC, MC, MCMF, MF, MFF, F	b–d	b–d
JPWSBP	WMW	C, CMC, MC, MCMF, MF, MFF, F	b–d	b–d
JPWSTA	WMW	C, CMC, MC, MCMF, MF, MFF, F	b–d	b–d
JPWSWB	WMW	C, CMC, MC, MCMF, MF, MFF, F	b–d	b–d
JPWS	MW	C, CMC, MC, MCMF, MF, MFF, F	d	d
JPWSBP	MW	C, CMC, MC, MCMF, MF, MFF, F	d	d
JPWSTA	MW	C, CMC, MC, MCMF, MF, MFF, F	d	d
JPWSWB	MW	C, CMC, MC, MCMF, MF, MFF, F	d	d
JPWS	MWI	C, CMC, MC, MCMF, MF, MFF, F	d–h	d–h
JPWSBP	MWI	C, CMC, MC, MCMF, MF, MFF, F	d–h	d–h
JPWSTA	MWI	C, CMC, MC, MCMF, MF, MFF, F	d–h	d–h
JPWSWB	MWI	C, CMC, MC, MCMF, MF, MFF, F	d–h	d–h
JPWS	I, IP, P, PVP	C, CMC, MC, MCMF, MF, MFF, F, O	h	h
JPWSBP	I, IP, P, PVP	C, CMC, MC, MCMF, MF, MFF, F, O	h	h
JPWSTA	I, IP, P, PVP	C, CMC, MC, MCMF, MF, MFF, F, O	h	h
JPWSWB	I, IP, P, PVP	C, CMC, MC, MCMF, MF, MFF, F, O	h	h
Aspen, wS and/or wB with tamarack				
TATL	WMW, MW	C, CMC, MC, MCMF, MF, MFF, F	e	e
TAWBTL	WMW, MW	C, CMC, MC, MCMF, MF, MFF, F	e	e
WBTATL	WMW, MW	C, CMC, MC, MCMF, MF, MFF, F	e	e
WBTL	WMW, MW	C, CMC, MC, MCMF, MF, MFF, F	e	e
WSTL	WMW, MW	C, CMC, MC, MCMF, MF, MFF, F	e	e
WSTLTA	WMW, MW	C, CMC, MC, MCMF, MF, MFF, F	e	e
TATL	MWI, I, IP, P	C, CMC, MC, MCMF, MF, MFF, F	h	h

	INVENTORY MAINTENANCE MAP DATA			
MAIN CANOPY TREE SPECIES	MAP DRAINAGE	MAP TEXTURE	ECOSITE GENERAL	ECOSITE SECONDARY
TAWBTL	MWI, I, IP, P	C, CMC, MC, MCMF, MF, MFF, F	h	h
WBTATL	MWI, I, IP, P	C, CMC, MC, MCMF, MF, MFF, F	h	h
WBTL	MWI, I, IP, P	C, CMC, MC, MCMF, MF, MFF, F	h	h
WSTL	MWI, I, IP, P	C, CMC, MC, MCMF, MF, MFF, F	h	h
WSTLTA	MWI, I, IP, P	C, CMC, MC, MCMF, MF, MFF, F	h	h
TATL	P, PVP	O	l	l
TAWBTL	P, PVP	O	l	l
WBTATL	P, PVP	O	l	l
WBTL	P, PVP	O	l	l
WSTL	P, PVP	O	l	l
WSTLTA	P, PVP	O	l	l
Jack pine and black spruce with tA, wB or bP as secondary species				
R, RW, W, WMW, MW	C, CMC, MC	c	c	c
JPBSTA	R, RW, W, WMW, MW	C, CMC, MC	c	c
JPBSWB	R, RW, W, WMW, MW	C, CMC, MC	c	c
JPBSBP	W, WMW, MW	MCMF, MF, MFF, F	d	d
JPBSTA	W, WMW, MW	MCMF, MF, MFF, F	d	d
JPBSWB	W, WMW, MW	MCMF, MF, MFF, F	d	d
JPBSBP	MWI	C, CMC, MC	c–g	c–g
JPBSTA	MWI	C, CMC, MC	c–g	c–g
JPBSWB	MWI	C, CMC, MC	c–g	c–g
JPBSBP	MWI	MCMF, MF, MFF, F	d–g	d–g
JPBSTA	MWI	MCMF, MF, MFF, F	d–g	d–g
JPBSWB	MWI	MCMF, MF, MFF, F	d–g	d–g
JPBSBP	I, IP, P	C, CMC, MC, MCMF, MF, MFF, F	g	g
JPBSTA	I, IP, P	C, CMC, MC, MCMF, MF, MFF, F	g	g
JPBSWB	I, IP, P	C, CMC, MC, MCMF, MF, MFF, F	g	g
JPBSBP	P, PVP	O	k	k
JPBSTA	P, PVP	O	k	k
JPBSWB	P, PVP	O	k	k
Black spruce and jack pine with tA, wB or bP as secondary species				
R, RW, W, WMW, MW	C, CMC, MC	c	c	c
BSJPTA	R, RW, W, WMW, MW	C, CMC, MC	c	c
BSJPWB	R, RW, W, WMW, MW	C, CMC, MC	c	c
BSJPBP	W, WMW, MW	MCMF, MF, MFF, F	d	d
BSJPTA	W, WMW, MW	MCMF, MF, MFF, F	d	d
BSJPWB	W, WMW, MW	MCMF, MF, MFF, F	d	d
BSJPBP	MWI	C, CMC, MC	c–g	c–g
BSJPTA	MWI	C, CMC, MC	c–g	c–g
BSJPWB	MWI	C, CMC, MC	c–g	c–g
BSJPBP	MWI	MCMF, MF, MFF, F	d–g	d–g
BSJPTA	MWI	MCMF, MF, MFF, F	d–g	d–g
BSJPWB	MWI	MCMF, MF, MFF, F	d–g	d–g
BSJPBP	I, IP, P	C, CMC, MC, MCMF, MF, MFF, F	g	g
BSJPTA	I, IP, P	C, CMC, MC, MCMF, MF, MFF, F	g	g
BSJPWB	I, IP, P	C, CMC, MC, MCMF, MF, MFF, F	g	g
BSJPBP	P, PVP	O	k	k
BSJPTA	P, PVP	O	k	k
BSJPWB	P, PVP	O	k	k
Balsam poplar mixtures				
BFBP	W, WMW	C, CMC, MC, MCMF, MF, MFF, F	d–e	f
BFWSBP	W, WMW	C, CMC, MC, MCMF, MF, MFF, F	d–e	f
BP	W, WMW	C, CMC, MC, MCMF, MF, MFF, F	d–e	f

	INVENTORY MAINTENANCE MAP DATA			
MAIN CANOPY TREE SPECIES	MAP DRAINAGE	MAP TEXTURE	ECOSITE GENERAL	ECOSITE SECONDARY
BPBF	W, WMW	C, CMC, MC, MCMF, MF, MFF, F	d–e	f
BPBS	W, WMW	C, CMC, MC, MCMF, MF, MFF, F	d–e	f
BPJP	W, WMW	C, CMC, MC, MCMF, MF, MFF, F	d–e	f
BPTA	W, WMW	C, CMC, MC, MCMF, MF, MFF, F	d–e	f
BPTABF	W, WMW	C, CMC, MC, MCMF, MF, MFF, F	d–e	f
BPTABS	W, WMW	C, CMC, MC, MCMF, MF, MFF, F	d–e	f
BPTAWS	W, WMW	C, CMC, MC, MCMF, MF, MFF, F	d–e	f
BPWB	W, WMW	C, CMC, MC, MCMF, MF, MFF, F	d–e	f
BPWBBF	W, WMW	C, CMC, MC, MCMF, MF, MFF, F	d–e	f
BPWBBS	W, WMW	C, CMC, MC, MCMF, MF, MFF, F	d–e	f
BPWBWS	W, WMW	C, CMC, MC, MCMF, MF, MFF, F	d–e	f
BPWS	W, WMW	C, CMC, MC, MCMF, MF, MFF, F	d–e	f
BSBFBP	W, WMW	C, CMC, MC, MCMF, MF, MFF, F	d–e	f
BSBP	W, WMW	C, CMC, MC, MCMF, MF, MFF, F	d–e	f
BSWSBP	W, WMW	C, CMC, MC, MCMF, MF, MFF, F	d–e	f
TABP	W, WMW	C, CMC, MC, MCMF, MF, MFF, F	d–e	f
TABPBF	W, WMW	C, CMC, MC, MCMF, MF, MFF, F	d–e	f
TABPBS	W, WMW	C, CMC, MC, MCMF, MF, MFF, F	d–e	f
TABPJP	W, WMW	C, CMC, MC, MCMF, MF, MFF, F	d–e	f
TABPWS	W, WMW	C, CMC, MC, MCMF, MF, MFF, F	d–e	f
WBBP	W, WMW	C, CMC, MC, MCMF, MF, MFF, F	d–e	f
WBBPBS	W, WMW	C, CMC, MC, MCMF, MF, MFF, F	d–e	f
WBBPJP	W, WMW	C, CMC, MC, MCMF, MF, MFF, F	d–e	f
WBBPWS	W, WMW	C, CMC, MC, MCMF, MF, MFF, F	d–e	f
WSBFBP	W, WMW	C, CMC, MC, MCMF, MF, MFF, F	d–e	f
WSBP	W, WMW	C, CMC, MC, MCMF, MF, MFF, F	d–e	f
WSBSBP	W, WMW	C, CMC, MC, MCMF, MF, MFF, F	d–e	f
BFBP	MW, MWI, I	C, CMC, MC, MCMF, MF, MFF, F	e	f
BFWSBP	MW, MWI, I	C, CMC, MC, MCMF, MF, MFF, F	e	f
BP	MW, MWI, I	C, CMC, MC, MCMF, MF, MFF, F	e	f
BPBF	MW, MWI, I	C, CMC, MC, MCMF, MF, MFF, F	e	f
BPBS	MW, MWI, I	C, CMC, MC, MCMF, MF, MFF, F	e	f
BPJP	MW, MWI, I	C, CMC, MC, MCMF, MF, MFF, F	e	f
BPTA	MW, MWI, I	C, CMC, MC, MCMF, MF, MFF, F	e	f
BPTABF	MW, MWI, I	C, CMC, MC, MCMF, MF, MFF, F	e	f
BPTABS	MW, MWI, I	C, CMC, MC, MCMF, MF, MFF, F	e	f
BPTAWS	MW, MWI, I	C, CMC, MC, MCMF, MF, MFF, F	e	f
BPWB	MW, MWI, I	C, CMC, MC, MCMF, MF, MFF, F	e	f
BPWBBF	MW, MWI, I	C, CMC, MC, MCMF, MF, MFF, F	e	f
BPWBBS	MW, MWI, I	C, CMC, MC, MCMF, MF, MFF, F	e	f
BPWBWS	MW, MWI, I	C, CMC, MC, MCMF, MF, MFF, F	e	f
BPWS	MW, MWI, I	C, CMC, MC, MCMF, MF, MFF, F	e	f
BSBFBP	MW, MWI, I	C, CMC, MC, MCMF, MF, MFF, F	e	f
BSBP	MW, MWI, I	C, CMC, MC, MCMF, MF, MFF, F	e	f
BSWSBP	MW, MWI, I	C, CMC, MC, MCMF, MF, MFF, F	e	f
TABP	MW, MWI, I	C, CMC, MC, MCMF, MF, MFF, F	e	f
TABPBF	MW, MWI, I	C, CMC, MC, MCMF, MF, MFF, F	e	f
TABPBS	MW, MWI, I	C, CMC, MC, MCMF, MF, MFF, F	e	f
TABPJP	MW, MWI, I	C, CMC, MC, MCMF, MF, MFF, F	e	f
TABPWS	MW, MWI, I	C, CMC, MC, MCMF, MF, MFF, F	e	f
WBBP	MW, MWI, I	C, CMC, MC, MCMF, MF, MFF, F	e	f
WBBPBS	MW, MWI, I	C, CMC, MC, MCMF, MF, MFF, F	e	f
WBBPJP	MW, MWI, I	C, CMC, MC, MCMF, MF, MFF, F	e	f
WBBPWS	MW, MWI, I	C, CMC, MC, MCMF, MF, MFF, F	e	f
WSBFBP	MW, MWI, I	C, CMC, MC, MCMF, MF, MFF, F	e	f
WSBP	MW, MWI, I	C, CMC, MC, MCMF, MF, MFF, F	e	f

	INVENTORY MAINTENANCE MAP DATA			
MAIN CANOPY TREE SPECIES	MAP DRAINAGE	MAP TEXTURE	ECOSITE GENERAL	ECOSITE SECONDARY
WSBSBP	MW, MWI, I	C, CMC, MC, MCMF, MF, MFF, F	e	f
BFBP	IP	C, CMC, MC, MCMF, MF, MFF, F	e–h	f
BFWSBP	IP	C, CMC, MC, MCMF, MF, MFF, F	e–h	f
BP	IP	C, CMC, MC, MCMF, MF, MFF, F	e–h	f
BPBF	IP	C, CMC, MC, MCMF, MF, MFF, F	e–h	f
BPBS	IP	C, CMC, MC, MCMF, MF, MFF, F	e–h	f
BPJP	IP	C, CMC, MC, MCMF, MF, MFF, F	e–h	f
BPTA	IP	C, CMC, MC, MCMF, MF, MFF, F	e–h	f
BPTABF	IP	C, CMC, MC, MCMF, MF, MFF, F	e–h	f
BPTABS	IP	C, CMC, MC, MCMF, MF, MFF, F	e–h	f
BPTAWS	IP	C, CMC, MC, MCMF, MF, MFF, F	e–h	f
BPWB	IP	C, CMC, MC, MCMF, MF, MFF, F	e–h	f
BPWBBF	IP	C, CMC, MC, MCMF, MF, MFF, F	e–h	f
BPWBBS	IP	C, CMC, MC, MCMF, MF, MFF, F	e–h	f
BPWBWS	IP	C, CMC, MC, MCMF, MF, MFF, F	e–h	f
BPWS	IP	C, CMC, MC, MCMF, MF, MFF, F	e–h	f
BSBFBP	IP	C, CMC, MC, MCMF, MF, MFF, F	e–h	f
BSBP	IP	C, CMC, MC, MCMF, MF, MFF, F	e–h	f
BSWSBP	IP	C, CMC, MC, MCMF, MF, MFF, F	e–h	f
TABP	IP	C, CMC, MC, MCMF, MF, MFF, F	e–h	f
TABPBF	IP	C, CMC, MC, MCMF, MF, MFF, F	e–h	f
TABPBS	IP	C, CMC, MC, MCMF, MF, MFF, F	e–h	f
TABPJP	IP	C, CMC, MC, MCMF, MF, MFF, F	e–h	f
TABPWS	IP	C, CMC, MC, MCMF, MF, MFF, F	e–h	f
WBBP	IP	C, CMC, MC, MCMF, MF, MFF, F	e–h	f
WBBPBS	IP	C, CMC, MC, MCMF, MF, MFF, F	e–h	f
WBBPJP	IP	C, CMC, MC, MCMF, MF, MFF, F	e–h	f
WBBPWS	IP	C, CMC, MC, MCMF, MF, MFF, F	e–h	f
WSBFBP	IP	C, CMC, MC, MCMF, MF, MFF, F	e–h	f
WSBP	IP	C, CMC, MC, MCMF, MF, MFF, F	e–h	f
WSBSBP	IP	C, CMC, MC, MCMF, MF, MFF, F	e–h	f
BFBP	P, PVP	C, CMC, MC, MCMF, MF, MFF, F, O	h	h
BFWSBP	P, PVP	C, CMC, MC, MCMF, MF, MFF, F, O	h	h
BP	P, PVP	C, CMC, MC, MCMF, MF, MFF, F, O	h	h
BPBF	P, PVP	C, CMC, MC, MCMF, MF, MFF, F, O	h	h
BPBS	P, PVP	C, CMC, MC, MCMF, MF, MFF, F, O	h	h
BPJP	P, PVP	C, CMC, MC, MCMF, MF, MFF, F, O	h	h
BPTA	P, PVP	C, CMC, MC, MCMF, MF, MFF, F, O	h	h
BPTABF	P, PVP	C, CMC, MC, MCMF, MF, MFF, F, O	h	h
BPTABS	P, PVP	C, CMC, MC, MCMF, MF, MFF, F, O	h	h
BPTAWS	P, PVP	C, CMC, MC, MCMF, MF, MFF, F, O	h	h
BPWB	P, PVP	C, CMC, MC, MCMF, MF, MFF, F, O	h	h
BPWBBF	P, PVP	C, CMC, MC, MCMF, MF, MFF, F, O	h	h
BPWBBS	P, PVP	C, CMC, MC, MCMF, MF, MFF, F, O	h	h
BPWBWS	P, PVP	C, CMC, MC, MCMF, MF, MFF, F, O	h	h
BPWS	P, PVP	C, CMC, MC, MCMF, MF, MFF, F, O	h	h
BSBFBP	P, PVP	C, CMC, MC, MCMF, MF, MFF, F, O	h	h
BSBP	P, PVP	C, CMC, MC, MCMF, MF, MFF, F, O	h	h
BSWSBP	P, PVP	C, CMC, MC, MCMF, MF, MFF, F, O	h	h
TABP	P, PVP	C, CMC, MC, MCMF, MF, MFF, F, O	h	h
TABPBF	P, PVP	C, CMC, MC, MCMF, MF, MFF, F, O	h	h
TABPBS	P, PVP	C, CMC, MC, MCMF, MF, MFF, F, O	h	h
TABPJP	P, PVP	C, CMC, MC, MCMF, MF, MFF, F, O	h	h
TABPWS	P, PVP	C, CMC, MC, MCMF, MF, MFF, F, O	h	h
WBBP	P, PVP	C, CMC, MC, MCMF, MF, MFF, F, O	h	h
WBBPBS	P, PVP	C, CMC, MC, MCMF, MF, MFF, F, O	h	h

	INVENTORY MAINTENANCE MAP DATA			
MAIN CANOPY TREE SPECIES	MAP DRAINAGE	MAP TEXTURE	ECOSITE GENERAL	ECOSITE SECONDARY
WBBPJP	P, PVP	C, CMC, MC, MCMF, MF, MFF, F, O	h	h
WBBPWS	P, PVP	C, CMC, MC, MCMF, MF, MFF, F, O	h	h
WSBFBP	P, PVP	C, CMC, MC, MCMF, MF, MFF, F, O	h	h
WSBP	P, PVP	C, CMC, MC, MCMF, MF, MFF, F, O	h	h
WSBSBP	P, PVP	C, CMC, MC, MCMF, MF, MFF, F, O	h	h
Balsam poplar mixtures w/tamarack				
BPTATL	W, WMW, MW	C, CMC, MC, MCMF, MF, MFF, F	e	f
BPTL	W, WMW, MW	C, CMC, MC, MCMF, MF, MFF, F	e	f
BPWBTL	W, WMW, MW	C, CMC, MC, MCMF, MF, MFF, F	e	f
TABPTL	W, WMW, MW	C, CMC, MC, MCMF, MF, MFF, F	e	f
WBBPTL	W, WMW, MW	C, CMC, MC, MCMF, MF, MFF, F	e	f
BPTATL	MWI, I, IP, P	C, CMC, MC, MCMF, MF, MFF, F	h	f
BPTL	MWI, I, IP, P	C, CMC, MC, MCMF, MF, MFF, F	h	f
BPWBTL	MWI, I, IP, P	C, CMC, MC, MCMF, MF, MFF, F	h	f
TABPTL	MWI, I, IP, P	C, CMC, MC, MCMF, MF, MFF, F	h	f
WBBPTL	MWI, I, IP, P	C, CMC, MC, MCMF, MF, MFF, F	h	f
BPTATL	P, PVP	C, CMC, MC, MCMF, MF, MFF, F, O	h	h
BPTL	P, PVP	C, CMC, MC, MCMF, MF, MFF, F, O	h	h
BPWBTL	P, PVP	C, CMC, MC, MCMF, MF, MFF, F, O	h	h
TABPTL	P, PVP	C, CMC, MC, MCMF, MF, MFF, F, O	h	h
WBBPTL	P, PVP	C, CMC, MC, MCMF, MF, MFF, F, O	h	h
Ash, maple and elm stands				
BPMM	W, WMW, MW, MWI, I, IP, P	C, CMC, MC, MCMF, MF, MFF, F	f	f
BPMMBF	W, WMW, MW, MWI, I, IP, P	C, CMC, MC, MCMF, MF, MFF, F	f	f
BPMMBS	W, WMW, MW, MWI, I, IP, P	C, CMC, MC, MCMF, MF, MFF, F	f	f
BPMMWS	W, WMW, MW, MWI, I, IP, P	C, CMC, MC, MCMF, MF, MFF, F	f	f
BPWE	W, WMW, MW, MWI, I, IP, P	C, CMC, MC, MCMF, MF, MFF, F	f	f
BPWEWS	W, WMW, MW, MWI, I, IP, P	C, CMC, MC, MCMF, MF, MFF, F	f	f
MM	W, WMW, MW, MWI, I, IP, P	C, CMC, MC, MCMF, MF, MFF, F	f	f
MMBP	W, WMW, MW, MWI, I, IP, P	C, CMC, MC, MCMF, MF, MFF, F	f	f
MMTA	W, WMW, MW, MWI, I, IP, P	C, CMC, MC, MCMF, MF, MFF, F	f	f
MMWB	W, WMW, MW, MWI, I, IP, P	C, CMC, MC, MCMF, MF, MFF, F	f	f
MMWBTL	W, WMW, MW, MWI, I, IP, P	C, CMC, MC, MCMF, MF, MFF, F	f	f
MMWBWS	W, WMW, MW, MWI, I, IP, P	C, CMC, MC, MCMF, MF, MFF, F	f	f
MMWE	W, WMW, MW, MWI, I, IP, P	C, CMC, MC, MCMF, MF, MFF, F	f	f
MMWS	W, WMW, MW, MWI, I, IP, P	C, CMC, MC, MCMF, MF, MFF, F	f	f
TAMM	W, WMW, MW, MWI, I, IP, P	C, CMC, MC, MCMF, MF, MFF, F	f	f
WBMM	W, WMW, MW, MWI, I, IP, P	C, CMC, MC, MCMF, MF, MFF, F	f	f
WBMMBS	W, WMW, MW, MWI, I, IP, P	C, CMC, MC, MCMF, MF, MFF, F	f	f
WE	W, WMW, MW, MWI, I, IP, P	C, CMC, MC, MCMF, MF, MFF, F	f	f
WEMM	W, WMW, MW, MWI, I, IP, P	C, CMC, MC, MCMF, MF, MFF, F	f	f
BPMM	P, PVP	O	h	h
BPMMBF	P, PVP	O	h	h
BPMMBS	P, PVP	O	h	h
BPMMWS	P, PVP	O	h	h
BPWE	P, PVP	O	h	h
BPWEWS	P, PVP	O	h	h
MM	P, PVP	O	h	h
MMBP	P, PVP	O	h	h
MMTA	P, PVP	O	h	h
MMWB	P, PVP	O	h	h
MMWBTL	P, PVP	O	h	h
MMWBWS	P, PVP	O	h	h
MMWE	P, PVP	O	h	h

	INVENTORY MAINTENANCE MAP DATA			
MAIN CANOPY TREE SPECIES	MAP DRAINAGE	MAP TEXTURE	ECOSITE GENERAL	ECOSITE SECONDARY
MMWS	P, PVP	O	h	h
TAMM	P, PVP	O	h	h
WBMM	P, PVP	O	h	h
WBMMBS	P, PVP	O	h	h
WE	P, PVP	O	h	h
WEMM	P, PVP	O	h	h
Black spruce and tamarack w/Manitoba maple				
BSTLMM	P, PVP	C, CMC, MC, MCMF, MF, MFF, F, O	k	k
Tamarack as a pure species				
TL	W, WMW, MW	C, CMC, MC, MCMF, MF, MFF, F	e	e
TL	MWI, I	C, CMC, MC, MCMF, MF, MFF, F	h	h
TL	IP, P, PVP	C, CMC, MC, MCMF, MF, MFF, F, O	l	l
Tamarack w/black spruce as a secondary species				
TLBS	W, WMW, MW	C, CMC, MC, MCMF, MF, MFF, F	e	e
TLBS	MWI, I	C, CMC, MC, MCMF, MF, MFF, F	h	h
TLBS	IP, P, PVP	C, CMC, MC, MCMF, MF, MFF, F, O	k	k
Tamarack w/bS, bP, tA, wB and/or jP				
TLBSBP	W, WMW, MW	C, CMC, MC, MCMF, MF, MFF, F	e	e
TLBSTA	W, WMW, MW	C, CMC, MC, MCMF, MF, MFF, F	e	e
TLBSWB	W, WMW, MW	C, CMC, MC, MCMF, MF, MFF, F	e	e
TLJP	W, WMW, MW	C, CMC, MC, MCMF, MF, MFF, F	e	e
TLBSBP	MWI, I	C, CMC, MC, MCMF, MF, MFF, F	h	h
TLBSTA	MWI, I	C, CMC, MC, MCMF, MF, MFF, F	h	h
TLBSWB	MWI, I	C, CMC, MC, MCMF, MF, MFF, F	h	h
TLJP	MWI, I	C, CMC, MC, MCMF, MF, MFF, F	h	h
TLBSBP	IP, P, PVP	C, CMC, MC, MCMF, MF, MFF, F, O	k	k
TLBSTA	IP, P, PVP	C, CMC, MC, MCMF, MF, MFF, F, O	k	k
TLBSWB	IP, P, PVP	C, CMC, MC, MCMF, MF, MFF, F, O	k	k
TLJP	IP, P, PVP	C, CMC, MC, MCMF, MF, MFF, F, O	k	k
Tamarack w/bP, wB, wS and/or tA				
TLBP	W, WMW, MW	C, CMC, MC, MCMF, MF, MFF, F	e	e
TLTA	W, WMW, MW	C, CMC, MC, MCMF, MF, MFF, F	e	e
TLWB	W, WMW, MW	C, CMC, MC, MCMF, MF, MFF, F	e	e
TLWS	W, WMW, MW	C, CMC, MC, MCMF, MF, MFF, F	e	e
TLWSWB	W, WMW, MW	C, CMC, MC, MCMF, MF, MFF, F	e	e
TLBP	MWI, I	C, CMC, MC, MCMF, MF, MFF, F	h	h
TLTA	MWI, I	C, CMC, MC, MCMF, MF, MFF, F	h	h
TLWB	MWI, I	C, CMC, MC, MCMF, MF, MFF, F	h	h
TLWS	MWI, I	C, CMC, MC, MCMF, MF, MFF, F	h	h
TLWSWB	MWI, I	C, CMC, MC, MCMF, MF, MFF, F	h	h
TLBP	IP, P, PVP	C, CMC, MC, MCMF, MF, MFF, F, O	l	l
TLTA	IP, P, PVP	C, CMC, MC, MCMF, MF, MFF, F, O	l	l
TLWB	IP, P, PVP	C, CMC, MC, MCMF, MF, MFF, F, O	l	l
TLWS	IP, P, PVP	C, CMC, MC, MCMF, MF, MFF, F, O	l	l
TLWSWB	IP, P, PVP	C, CMC, MC, MCMF, MF, MFF, F, O	l	l

[a] Lindenas, D.G. 1985. Specifications for the interpretation and mapping of aerial photographs in the forest inventory section. Saskatchewan Parks and Renew. Resourc. For. Div., Prince Albert, Saskatchewan. Intern. Doc.

[b] Beckingham, J.D.; Nielsen, D.G.; Futoransky, V.A. 1996. Field guide to ecosites of the mid-boreal ecoregions of Saskatchewan. Nat. Resour. Can., Can. For. Serv., Northwest Reg., North. For. Cent., Edmonton, Alberta. Spec. Rep. 6.